The Eating Instinct

The Eating Instinct

Food Culture, Body Image,
and Guilt in America

Virginia Sole-Smith

Henry Holt and Company   New York

Henry Holt and Company
Publishers since 1866
175 Fifth Avenue
New York, New York 10010
www.henryholt.com

Henry Holt® and 🔲® are registered trademarks of
Macmillan Publishing Group, LLC.

Library of Congress Cataloging-in-Publication Data

Names: Sole-Smith, Virginia, author.
Title: The eating instinct : food culture, body image, and guilt in America /
    by Virginia Sole-Smith.
Description: First edition. | New York : Henry Holt and Company, 2018. |
    Includes bibliographical references and index.
Identifiers: LCCN 2018013104 | ISBN 9781250120984 (hardcover)
Subjects: LCSH: Food habits. | Body image.
Classification: LCC GT2850 .S627 2018 | DDC 394.1/2—dc23
LC record available at https://lccn.loc.gov/2018013104

Our books may be purchased in bulk for promotional, educational, or business use. Please
contact your local bookseller or the Macmillan Corporate and Premium Sales Department at
(800) 221-7945, extension 5442, or by e-mail at MacmillanSpecialMarkets@macmillan.com.

First Edition 2018

Designed by Kelly S. Too

Printed in the United States of America

1   3   5   7   9   10   8   6   4   2

For Violet, who led the way.

CONTENTS

PREFACE

What does it mean to learn to eat, in a world that's constantly telling us not to eat? It's a question I started asking five years ago, when my daughter Violet stopped eating as a result of severe medical trauma. Suddenly, we had to begin again, to forget all the normal rules about breast-feeding and spoon-feeding, and gingerly pick our way through a surreal new world where food was simultaneously the enemy and our salvation. But in many ways, this is also a question I've been asking my whole life, as a woman who came of age at the intersection of the alternative-food movement and the war on obesity. As a skinny kid growing up in the 1980s, I thought processed foods were great; I felt sorry for my friends whose moms bought only weird brown bread for their peanut-butter-and-jelly sandwiches. I'm not saying we never thought about healthy eating—the 1970s and 1980s also saw the birth of modern diet culture, with the rise of aerobics videos and fat-free everything. And

I certainly understood that fat was bad, and that was why we bought skim milk and diet soda. But this was a more straight-forward time for dieting; you joined Weight Watchers and ate SnackWell's if you needed to get thin. You didn't have to reject an entire food-industrial complex or introduce exotic new ingredients into your diet. Quinoa was still relegated to the dusty bin in a corner of our town's one hippie-run health food store.

But by the time I graduated from high school in 1999, we were buying mesclun greens and whole-grain pasta. Obesity had become an official public health crisis. Carbohydrates were the new "bad food," though fat was far from vindicated. We were still a few years away from the landmark publication of Michael Pollan's *The Omnivore's Dilemma*, but the conversation was beginning—on the coasts, at least—about the importance of organic farming and the need to eat "whole foods" instead of processed ones. As I'll explore in the chapters ahead, these twin anxieties about obesity and about the eco-health implications of our modern food system have transformed American food and diet culture. Eating well has become wildly more complicated; it's now about "eating clean," it's about being a socially responsible consumer and an accomplished home chef. Thinness has become our main measure of health, but also of personal virtue, of having the right kind of education, politics, and morality.

When I went to college, I got homesick, found a website that would deliver Ben & Jerry's to my dorm room, and gained a bit more than the freshman fifteen. It turned out that being a thin child did not ensure that I would be a thin adult; I've spent most of my twenties and thirties perpetually in the process of losing or gaining the same forty pounds. I'm often not even aware that it's happening; I'll just suddenly find that, yet again, my jeans fit all wrong in one way or another. Most of the time, I live firmly in the "overweight" range of the body mass index scale. I've never been what Roxane Gay calls Lane Bryant fat, which I say not as any point of pride, but just to be clear that I'm not here to appropriate the experience of anyone who has lived in a larger body and knows firsthand the daily discrimination that brings. But my body did go from thin, to normal, to heavy-ish during a fifteen-year period when our whole culture got a lot more anxious about food and weight, when the ideal woman's body went from merely thin, to thin and impossibly toned, capable of running marathons, pretzeling into complex yoga positions, and breast-feeding a baby all at the same time. Of course those messages have seeped into my psyche; of course they have shaped how I feel after I eat a doughnut or a salad. I've never had an eating disorder or even much of a propensity toward dieting. I lasted exactly two days on the one and only crash diet I tried, before running to the deli for a giant turkey sandwich. But I am not always at peace with my body. After all, I've never met anyone who really is.

You may be wondering what my weight and dieting history has to do with having a baby who won't eat. But nothing connected with food happens in a vacuum. Modern diet culture doesn't just happen to teenage girls trying to lose weight for prom. It's influencing how all of us think about food, every day and at every meal, often in hidden and unconscious ways. This culture tells pregnant women we have to eat a certain way, and then feed our babies and young children according to unattainable standards of perfection. It draws the lines between which sorts of eating habits are normal, acceptable, and healthy, and which are unhealthy, even disordered and pathological. And it targets both men and women, though we'll hear a bit more from women in this book, because we are still the ones held to the strictest weight and beauty standards, suffering from disproportionately higher rates of disordered eating, and feeling the most pressure around food and parenting. But diet trends do not discriminate on the basis of age, weight, race, or social class; they target each of these groups in specific and personal ways.

Recognizing when my ideas about how I should feed Violet were the product of external pressures helped me navigate the process of teaching her to eat again. I'm not sure I would have learned any of this if we hadn't had to start all over, to figure out how to make food feel safe to a traumatized child. I'm not sure I'm done learning it now. Violet taught me that eating well cannot be about following rules; it has to be about trusting our own instincts, which value safety, comfort, and pleasure just as much as nutrition, and sometimes more. But in many ways, what we could help her overcome as a baby was only possible because she was still so young, and still very much insulated from the larger food world around us.

After I wrote a piece for *The New York Times Magazine* about how Violet learned to eat again, I was inundated with emails and comments from people struggling to figure out the same thing. I began talking to them and collecting their stories, some of which you'll read in the next chapters. When you consider our most intimate physical activities, eating is somewhere just below sex, showering, and using the toilet. Yet we do it in public, all the time. By agreeing to share their stories in this book, my sources are now eating on an even larger stage. I'm grateful for their honesty; in return, I've changed names and omitted or very slightly altered personal details when it was necessary to respect their privacy. I've also respected their preferences with regard to sharing certain facts, such as weight or specific eating choices. For some people, these are empowering truths to own and share. For others, to reveal weight, in particular, is emotionally destructive. (It can also be fraught to read about someone else's weight, so if you struggle with an eating disorder yourself, please use your own best judgment in deciding whether to read those chapters.)

Over the past two years, I've sat in immaculate marble kitchens and tiny, peeling-linoleum ones. I've visited doctors' offices, research labs, and commercial kitchens. I've interviewed some three dozen people about their relationship with food, and surveyed or exchanged emails with many more. And I've eaten many meals— with women recovering from weight-loss surgery or eating disorders; with picky eaters and adventurous gourmands; with people who have seemingly unlimited budgets and with folks struggling to afford to keep food in the house. Their individual struggles are unique and influenced by their own particular biology, family dynamics, socioeconomic status, and idiosyncratic tastes and preferences. The resulting myriad of ways they experience food is sometimes hard to fathom if you haven't ever met someone who grocery shops with food stamps, or who eats only french fries. But they're all products of our modern food culture. And they're asking the same questions that most of us have: How did I learn to eat this way? Why is it so hard to feel good about food? And how can I make it better?

The Eating Instinct

One

Nothing by Mouth

September 17, 2013. It is the day before my daughter Violet's one-month birthday. It is also the first day that she will almost die.

It is a day I'll always pause on, in the years to come. Mostly I'll remember strange, flash-frozen details: The receptionist's stricken face in our pediatrician's office as paramedics bustle us out to an ambulance. A respiratory technician in the emergency room shouting that he can't go help another patient because *I have a very sick little girl here.* Trying to catch up to the clatter of heels running ahead of us down a long hospital hallway as they race our daughter's crib away. They said we'd have time to kiss Violet good-bye before they took her to the operating room. I'm mad that they're rushing. I don't understand that it's because, suddenly, there is no time.

But sometimes I'll remember something else about this day, and it's the thing that's most important to the story you're reading now. September 17, 2013, is also the day when Violet stops eating.

And she won't start again, not in any meaningful way, for almost two years.

I don't realize any of this when Dan and I wake up a little after seven a.m., blinking and disoriented from an unexpected six-hour stretch of sleep. I groggily remember going into Violet's room when she cried sometime after midnight. I put her to my breast, but she was asleep again before she could latch. I put her back down; I was asleep again before I could think about when she'd last eaten, or whether that was normal for a newborn. We are all so tired. "This is having a new baby," we think.

Now, Violet is still sleeping, so I pump to relieve the pressure in my breasts and we stumble around, making coffee and breakfast. We have her four-week well-baby visit this morning and we need to hurry because we overslept, not expecting our newborn to let us do that. When she wakes up, Dan brings her to me to feed at the dining room table. She latches for a minute, and then pulls off, frustrated and sleepy again. "Is it weird she isn't eating?" I ask Dan. "She slept so long, shouldn't she be starving?" We don't know. We don't really worry. Everything has been fine. This must be fine, too.

Right from birth, Violet cried when she was hungry, slept when she was full. When awake, she stares at us intensely. When she sleeps, she sleeps a lot—one night, a few days ago, for nine hours. That next morning, I texted Michelle, the nurse practitioner we see at our pediatrician's office, because she is also my friend from yoga. She is a peaceful, comforting presence, the perfect ballast to what I know will be my paranoid new-mom tendencies. "You just got lucky!" she assured me. "What a great baby!" my friends say. Their children were fierce with colic and rage in the newborn months, and they are jealous that I'm getting to sleep. So I don't worry. Breast-feeding is fumbling and slow; Violet's latch hurts, a pain I feel all the way down to my toes. I use a nipple shield even though the lactation consultant tells me not to get dependent on it; when I go to her nursing circle, she stands over us and jams Violet's head onto my breast, and it works. "You're getting it," she says. "Keep in touch." I feel relieved when Michelle tells us that Violet has regained her birth weight at her two-week checkup. The time we're spending on breast-feeding sessions starts to drop, from forty-five minutes to twenty, then ten, then five; I think the two of us are just getting better at it.

We aren't.

This morning, when we finally stumble in to the four-week well visit, a nurse puts Violet on the baby scale and we see that she has somehow lost half a pound. She once again weighs less than the seven pounds, nine ounces she measured at birth. Michelle comes in laughing. "That can't be right!" But she weighs Violet again and it is. Then she inspects the purplish tinge of Violet's lips, feet, and fingernails. I think she's cold, but I have a hat in the diaper bag; I am prepared. Michelle lets me get the hat, while she calls downstairs for a cardiologist. I ask why. "She's dusky," Michelle says. She is still calm, but she doesn't lie to me. "That means something is wrong." She knows that Violet's heart is failing.

After that, information comes more quickly than I can grasp. The oxygen level in Violet's blood is only 75 percent of what it should be. By the time we're in the ambulance, lights flashing as it heads down the Taconic Parkway toward Maria Fareri Children's Hospital in Valhalla, New York, it's at 60 percent. I report this to my stepmother when she calls my cell phone; I am not worried, because the EMTs are relentlessly cheerful and making jokes about how we keep getting to pass cops. Mary tries to sound not worried for me, but I can tell she is scared. She worked in hospitals for years as a physician assistant. She has seen people die when their oxygen dropped so low. I don't know yet that every red blood cell in your body must have an oxygen molecule securely attached;

that when less than 80 percent of them do, your body starts to
shut down. That Violet's body is shutting down. In the pediatric
intensive care unit, Violet is put on a ventilator, the breathing tube
snaking down her throat before she's fully sedated. Even with
the machine breathing for her, her oxygen keeps dropping. By
the time they are running her away from us, she's at 20 percent.
In the catheterization lab, a cardiologist threads a catheter into
Violet's heart, inflates a tiny balloon, and tugs, punching a hole
through her interatrial septum to release a gush of pent-up oxy-
genated blood. This is the first time we break Violet to save her.

The next day, we begin to learn how several rare congenital
defects have made Violet's heart "incompatible with life," as one
doctor gently puts it. Violet is missing her mitral valve and part
of her ventricular septum; her left ventricle is smaller than it should
be, and her aorta and pulmonary artery have traded places. These
kinds of problems are known collectively as single ventricle physiology. They cannot be repaired, but a cardiothoracic surgeon can
cut apart veins and arteries and sew them back together in a life-
sustaining pattern over the course of three open-heart surgeries.
When the process succeeds, Fontan circulation, as the result is
known, enables a child to reach a healthy, if heavily monitored,
adulthood.

What the doctors cannot tell us, what we can't begin to grasp
as we sit in the PICU on Violet's one-month birthday, waiting for
her to breathe on her own, is that as difficult as the cardiac diagnosis is—and always will be—it won't be what dominates our daily
lives in the months ahead. Violet not eating is merely collateral
damage, a side effect of this much bigger, all-consuming problem.
But it will consume us, especially me. Her feeding tube will be a
constant reminder of this day and all the hard days ahead. It will
tell the world that we have a sick child.

Eating is fundamental to human existence. It's the primary

work of mothers and babies; the basis of every holiday and communal celebration; the first thing most of us do when we wake up in the morning. In the hospital, a patient who cannot (or must not) eat is referred to as "NPO"; the abbreviation stands for *nil per os*, Latin for "nothing by mouth." A child who takes nothing by mouth isn't participating fully in human life. It makes people wonder what else she can't do; it focuses us on her limits, instead of her potential.

I worry, on that first day and the next, in an idle, abstract way, about whether Violet is hungry. I understand intellectually that she's on a ventilator, that feeding is impossible, but my heart hasn't caught up yet and I keep thinking, "Aren't I supposed to feed her? Is this hospital not very pro-breast-feeding?" It doesn't register that my breasts don't hurt. By the time they race Violet away from us, it has been almost ten hours since I pumped and tried to feed her. I should be engorged, leaking. But I'm not. My body already knows that it is not needed. That my baby has stopped eating.

At some point, while we're waiting for Violet to live or die, a nurse brings me a breast pump and a bag of clear plastic breast-milk containers. I am intimidated by how many there are; a dozen two-ounce vials with snap tops. "Maybe I'll pump tomorrow," I think. "I'd rather just feed her directly," I tell the nurse. She kindly explains that Violet can't eat right now because she doesn't have the energy. This is also what had been happening during those shortening feeds over the past two weeks—she was too oxygen deprived to suck, swallow, and breathe, the trifecta of skills essential to infant feeding.

Once she's been stabilized, our doctors are determined that Violet regain the half pound lost while she was dying, so she's given a feeding tube almost as soon as she comes off the ventilator. The nurse inserts a nasogastric tube into Violet's nose, and then pushes it down her esophagus and into her stomach. The tube is connected to a feeding pump beside Violet's clear plastic bassinet. I begin pumping as well, cloistered in a windowless room down the hall that the PICU reserves for this purpose. The hospital's breast pump is mint green and waist-high. Its analog dials look like eyes. When it stares at me, I think of wacky robot characters on failed 1980s sitcoms. Back in our hospital room, I hand over my half-filled vials apologetically. The nurses must mix them with formula to make enough food for Violet. They run this cocktail through the feeding tube every three hours, pushing as many calories as possible into Violet whether she's awake or asleep.

Violet undergoes her first open-heart surgery a week later. This time, I'm not the only one to assume that breast-feeding will resume soon after Violet comes off her second ventilator. Everyone expects it. But she continues to tire so quickly that the doctors figure she's burning more calories trying to eat than she can possibly take in. So a second nasogastric tube goes in, this time with very little deliberation. We're calling it the NG tube now. We know what "NPO" means. We're getting the lingo down. But: "It's a temporary measure," they assure us. "Just till she gets her strength back." We think she will be eating normally within two weeks.

Instead, we are discharged after twenty-two days in the hospital and go home with the NG tube still in place. Before leaving the hospital, we run Violet's noon tube feed under the proud supervision of the nurses who have spent the past week training us to use the equipment just the way they do. Then I dress Violet in a purple onesie patterned with pink hearts. I bought it especially for our hospital departure day, paying too much for it while shopping online during one of my many pumping sessions. She screams when I pull the soft cotton over her head; we haven't figured out yet that anything near Violet's face terrifies her and that we need to switch to outfits with zippers or snaps. We tuck Violet into her car seat, which Dan carries into the elevator, swinging it gently to make her smile. In the hospital driveway, as we transfer the car seat into the back of our secondhand Subaru, Violet throws up all of the formula she has just been fed. The car is already packed and running; we are desperate to go home. We clean her off, throw the soaked onesie into a plastic grocery bag, and drive.

But now we have a new problem. Tube feeding makes Violet vomit. For most of the next year, she will throw up four to seven times a day. "Babies spit up," we are told. But this is something more. The vomit can slip out silently, or it can choke Violet, turning her little face purple with the effort of getting it out. For an hour after every tube feed, we are afraid to hold her in case the slightest jostling triggers her reflux. She spends most of her day swaddled and frozen in her bouncy chair or crib. We spend our nights with two baby monitors on our pillows, one of us leaping out of bed as soon as we hear the first cough. The chronic vomiting makes us even more desperate for Violet to eat by mouth—and it also makes that goal feel even less attainable. So every three hours, we circle through the same dance: First I try to get Violet to nurse. Next, Dan offers a bottle. Every three hours, we try bottle or breast, all the while taking detailed notes to record how long she latches on, or how many milliliters she swallows. It's never more than a teaspoonful. Every three hours, we try this most fundamental act of parenthood. And fail.

The day I stop breast-feeding is a rainy Saturday in early November. Violet is now nine weeks old. It's a quiet morning; I shuffle around in my glasses and yoga pants, the obligatory new-mom uniform. Violet plays on an old red quilt, batting at a stuffed tomato on her play gym with ferocity while I fold the laundry. She coos whenever I make faces at her. I sing songs and kiss her toes. Other mothers have told me how maternity leave can drag: all those endless days alone with a newborn. But our morning passes quickly, maybe because we've been alone so rarely in her short life, half of

which has now been spent in a hospital. Or maybe it's because the
doctors have emphasized the importance of adhering to Violet's
feeding schedule; she must be fed every three hours, because she's
still recovering from that first heart surgery and every calorie is critical. The breaks between feedings feel too short for me to accomplish more than one item on my To Do list. At 11:55 a.m., with
half the laundry still crumpled in its basket, I set up my nursing
pillow, unfasten my nursing bra, and make sure I have the stop-
watch on my iPhone ready to time.

Violet grins when I pick her up. But as soon as I turn her into
position—the "football hold" that I've practiced under the supervision of lactation consultants, feeding specialists, and nurses—her
little face changes. She begins to scrabble against me, yelling and
turning her head away to push her face into the friendly woodland
print on the pillow. I change to a different hold. I shift the angle of
her head, my elbow, my shoulder. I sing more and I talk to Violet,
explaining that she's okay, that this eating business is supposed
to be a good thing. Then I start to cry too. I cry because it isn't
working, and everyone said it would be by now. I cry because Violet's refusal to eat is so opaque, so absolute, that it defies adult
logic and explanation. And I cry because when I start to match
her desperation, when I start to want to kick and scrabble back,
instead I hold her head and force my breast into her mewling little
mouth, until she shrieks louder and begins to gag and sputter. I
remember the lactation consultant jamming Violet's tiny head into
place. I want to think that this is the same, that this is normal.
But I know my heart is racing too hard and I'm feeling a dangerous kind of fury.

And so I stop. I put Violet in her bouncy seat and bounce her
until she is quiet. "I'm sorry, I'm sorry, I'm sorry," I say as I bounce,
until we are both calm and slightly hypnotized. "We're done. We're
done now."

Then I take the long, skinny tube that dangles out of my baby's nose and connect it to a purple plastic feeding pump mounted next to her bouncy seat, which I've already loaded with precisely 84 milliliters of formula-fortified breast milk. The pump beeps and whirs, Violet drifts off to sleep, and I sit on the floor of my living room while, once again, a machine feeds my child.

Because babies begin nursing in the first hours of life, because the cry of hunger is one of our first communications with the world, it's widely assumed that eating is our most primitive instinct. This is not quite correct. Breathing is the first thing we do after birth, and perhaps the only behavior more fundamental to our survival. But hunger is instinctual. And so is satiation. We are born knowing to cry, root, and seek when we feel hunger, and to stop when we are full. But while these behaviors are innate, they are also surprisingly fragile, in need of constant reinforcement. A baby cries, a breast or bottle is offered, and the baby sucks and swallows until she feels better. Most newborns do little else in their first few months, until their ability to eat is finely honed and the feeding relationship between parent and child is thoroughly established. In this way, the instinct to eat isn't just a need for physical nourishment—it also ensures that babies form secure attachments. It's how they fall in love.

As a baby grows, the act of eating becomes more intricate. Chewing solid foods, using a spoon, and drinking from an open cup all require a complicated interplay of fine motor skills, gross motor skills, physical strength, stamina, and constant practice. For most babies, the process happens so seamlessly that the learning seems intuitive: the infant happily gums toys, then graduates to slurping down spoonfuls of applesauce, and soon associates such foods with the same satiety as she experienced through milk feeding. "Babies come into the world predisposed to learn all kinds of different things," says Leann Birch, a psychologist who studies infant feeding and childhood obesity at the University of Georgia. "There is a developmental timetable for when a baby can swallow food or move her tongue and jaws in certain ways. But without the right learning experiences, it won't come together."

And those learning experiences are surprisingly easy to miss. Only around 100,000 children in the United States have problems severe enough to require the use of a feeding tube, according to estimates by the Feeding Tube Awareness Foundation. But 25 percent to 45 percent of all children develop the kinds of habits that pediatricians and therapists see as the hallmarks of a "problem feeder." They may refuse to eat certain flavors, textures, or even entire food groups; others may eat too much, hoard food, or demand snacks at times when they can't possibly be hungry. Some degree of problem feeding is an expected phase of childhood. Colic, reflux, or a poor latch can cause an otherwise healthy infant to go on a temporary hunger strike. And virtually every toddler develops an increased suspicion of new foods. Some parents and pediatricians may panic over these normal developmental stages, while others may dismiss a sensory processing problem or weak oral motor skills as just picky eating.

Either reaction can result in a daily override of instinct for the finicky toddler whose parents turn every meal into a battle over "just one more bite," and conversely, for the stocky five-year-old whose worried parents ban treats and second helpings. Over time, a child can be conditioned by parental instructions to ignore her own instincts about how much to eat, but that conditioning doesn't always have the desired result. Studies have found that when children are rewarded for eating healthy foods, they tend to like those foods less and crave sweet treats more. "There's a tension here, because we need children to become socialized to eat at mealtimes,"

Birch acknowledges. "And yet parents often think they need to take more control of this than they should."

For Violet, learning not to eat was a quick and brutal process. Her eating instinct was destroyed almost as soon as it emerged, by what's known medically as an oral aversion. Also historically referred to as oral defensiveness, and more unnervingly as infantile anorexia, this condition results when a child refuses to eat as a way of protecting herself from perceived trauma. The diagnosis didn't resonate with us immediately. Violet's first smile happened in the hospital the day before her heart surgery; that didn't seem like a milestone that a traumatized baby would meet. And when we weren't trying to feed her, she was happy, alert, and curious about the world. But for Violet, not eating was, at its core, an act of survival. When tubes and her own failing heart compromised her ability to breathe, she fought off anything else that threatened that even more primal need. When the tube feeds caused her to vomit up formula and breast milk, she began to recoil at anything that smelled similar. And as a result of those early nursing struggles, the emergency intubation in the hospital, or, perhaps, our own ceaseless efforts to get her to eat, Violet forged a connection between eating and pain, just as dogs learned to salivate at the sound of a bell in Ivan Pavlov's classic experiment on conditioning. A baby with an oral aversion can lose those digestive reflexes and instead feel nauseated at the sight of a breast or bottle, or as soon as her feeding tube delivers formula, untasted, straight to her stomach.

Not eating was how Violet felt safe. In those first months, that felt bizarre, even wrong—how could any baby reject eating? But the truth is that Violet's story, while dramatic, is also profoundly normal. Not eating is something that many of us use to feel safe. So is overeating, though that safety is often short-lived and accompanied by guilt, especially when eating contributes to a body that is bigger than our world will accept. And so we are all of us learning not to eat, all the time, whenever we start a diet, banish gluten, or regretfully unbutton our too-tight jeans after a big meal. For many of us, not eating can be about safety and survival, but also about punishment: an atonement for past gluttony. It can be about a moral code: the belief that eating meat is an act of murder, or that organic farming will save the planet. Our periods of not eating are often followed by days of intensive eating, and then once again, the need to repent. Or we combine eating and not eating in even more nuanced and sometimes bizarre ways: We don't eat after certain times or on certain days. We eat protein but not carbohydrates. We'll chew our calories, but never drink them. We choose whole foods, not processed foods; butter, not margarine; breast milk, never formula.

Successful eating requires both a strong connection to those instincts and the right set of learned behaviors. Many children first unconsciously learn the importance of not eating from their parents. The lesson is reinforced by friends, teachers, and doctors, and, much sooner than you might think, by diet culture—the multibillion-dollar complex of diet books, magazines, food marketers, blogs, and social-media influencers who trade on convincing us that this food is good, but that one is bad. Even if they get through the first few years intact, by preschool most kids have begun to stop connecting to their internal sense of hunger and satiety, and instead rely on external cues to decide whether and how much to eat. Their teachers might make rules about the need to finish sandwiches before starting cookies at lunchtime. Grandparents praise "good eaters" while offering unlimited access to a snack cabinet. Dentists talk about "sugar bugs" and the need to avoid cavity-causing candy. Food marketers teach them that the best cereal boxes contain prizes. And this is all before kids enter kindergarten.

Really, when you consider the onslaught, it's remarkable that any of us ever manage to eat only when we're truly hungry, or stop when we're full. And when eating goes off the rails—as it has not only for the quarter of children diagnosed with feeding problems, but also for the 40 percent of teenage girls using "restrictive measures" to lose weight, and for many of the 36.5 percent of Americans struggling with obesity—we reach a crossroads: Do you try to correct the behavior—learning to eat according to a prescribed set of external rules and conditions—or do you try to rediscover those internal cues that tell us when to eat and when to stop? It's a divisive question among doctors, therapists, and anyone who studies food. And it's an intensely personal debate unfolding, explicitly or not, around most kitchen tables in the country.

Mothers, in particular, face incredible pressure to feed themselves and their kids perfectly. The last decade has seen a renewed public health obsession with breast-feeding, as well as the rise of the alternative-food movement and the "clean eating" trend, with its penchant for organic juice cleanses, detoxing, and whole-food diets. These forces have combined to raise the stakes enormously around how pregnant women and small children should eat. And there's the inevitable fact that, from the moment of conception, a mother's body and her child's health and happiness become all tangled together. When I was pregnant with Violet, I felt as if everything I did with my body mattered desperately, because my body was building the baby. And it seemed only logical that if I built her out of hormone-free dairy, organic vegetables, and three yoga classes a week, she would turn out better than if I built her out of the Cheez-Its and ginger ale that were all I really wanted to eat for twenty weeks straight. If I went to four yoga classes a week and walked three miles a day, that would be even better for her. If I gained too much weight, that would be worse. So I went to acupuncture, got massages, refused to take as much as a single Advil when I got a migraine.

And because I am human, I also broke some rules. I drank some wine on vacation. I ate fast food on a road trip. I lay on my couch and watched *West Wing* reruns instead of doing my prenatal core workouts. I told myself that it wouldn't matter, that these rules were designed to drive women crazy by holding us to unrealistic standards, just as any overly prescriptive diet does. And I mostly believed that, until my body built a baby who was smart, beautiful—and broken. I didn't know if it was the wine, the fast food, or some other unknown rule I hadn't followed. The doctors told me, over and over, that heart defects are caused by random genetic mutations far beyond my control. It didn't matter. I knew my body had failed—and that meant I had failed my child, long before I gave birth to her.

So when Violet stops eating, it is my second failure as her mother. My body has spent nine months getting ready to do this thing that every doctor, nurse, parenting website, and random lady on Facebook seems to agree is absolutely critical for her health and success in life. Then, suddenly, I can't do it.

Violet's feeding disorder started in a hospital, triggered by forces much bigger than either of us. But not every baby in such a situation stops eating. Once Violet recovers from her first heart surgery, there is no physical explanation for why she won't feed from breast or bottle. Her doctors are baffled to hear that she's still relying on the feeding tube for all her nutrition after we've been home for ten days, and then three weeks, and then five months. But she doesn't want to eat. She seems terrified by it. We're told that a mother's milk is the most comforting, healthiest, and most perfect food for a baby; that just the scent of it is soothing, because it means her mother is near. But my daughter finds no comfort in my milk. I worry that she finds no comfort in me.

I begin to search for answers every night at two a.m. when I'm awake and strapped into the breast pump. Even after we've stopped nursing, I keep pumping, because I'm sure this tiny baby still needs every bit of breast milk I can give her. I have to set my alarm for these pumping sessions, because on the feeding tube, Violet no longer wakes hungry during the night. Friends with healthy babies seem envious when they hear this; they don't realize that I would happily cuddle my daughter in the middle of the night; that instead, I clench my teeth against the mechanical tug of one robotic pump, while another whirs and pushes calories into Violet's NG tube. I stare into the blue glow of my laptop screen and Google "babies who won't eat," while these machines stand in for how we were supposed to do this. I pump for forty-five minutes every three hours, far longer than the lactation consultants advise, but I still only get, at best, two ounces of milk. My supply is dropping. The pumps break, clog, and beep. Technology has not perfected its replication of the process of nursing.

In these late-night research sessions, I learn that pediatric feeding therapy is often done by speech-language pathologists, because good oral motor skills are needed both to speak and to eat, or by behavioral psychologists who design highly structured training programs. (Physical therapists, occupational therapists, and pediatricians may also do this work; in most cases a family's choice is dictated by geographic availability.) The most common training approach, used by almost all of the nearly thirty feeding programs found in children's hospitals and private clinics around the country, is a ritualistic method known as one-to-one reinforcement. It's a form of behavior modification, a psychological tactic in which food refusal is classified as a negative behavior to be systematically replaced with a positive one. "Babies can grasp cause and effect very early," says Amy Kathryn Drayton, a leading behaviorist who directs the feeding program at the University of Michigan's C. S. Mott Children's Hospital. "I've seen children who learned to vomit at the sight of the bottle and an eight-month-old who could fake-cough because he knew that would make the feeding stop."

I talk to Drayton in late 2014, when I begin the process of turning all that late-night research into an article for *The New York Times Magazine* about babies who won't eat. As part of my reporting, I also visit the Pediatric Feeding and Swallowing Center at the Children's Hospital of Philadelphia, where Colleen Lukens, one of the program's behavioral psychologists, invites me to sit in an observation room. We watch through a two-way mirror as Ivy, a two-year-old recovering from a stroke, is spoon-fed her lunch by a clinical feeding specialist named Julianne Quenzer. A bin of toys sits at their feet; whenever Ivy swallows some puréed broccoli, Quenzer pulls out a fire truck and thrusts it at her: "Good job, Ivy! Way to swallow your bite!" Then she takes the toy back and offers more food. When Ivy spits out a bite, Quenzer scoops it up and replaces it in the little girl's mouth. "You've got to keep it in, Ivy!" When Ivy spits it out once more, the process repeats. Ivy swallows on the fifth try, and the toy reappears. "Good job, Ivy!"

The behavioral model presumes that children who don't eat need external motivations. Drayton and Lukens don't deny the existence of internal cues about hunger and fullness, but they say that many of these children are no longer responding to them. "The early tube-feeding experience often disrupts all of that," Drayton says. "Every inch of their being says, 'Stop eating, stop eating, stop eating.' If we let these children make all of their own choices, they would make bad choices. That's why we don't let two-year-olds get their own apartments."

When I call Lukens for an update four weeks after my visit, she tells me that Ivy has been discharged eating 3.5 ounces of puréed food and drinking four ounces of milk, three times daily. Nearly all the children who come through the program achieve their feeding goals by the month's end, but the hospital does not track long-term outcomes. Families often struggle to maintain the rigid feeding schedule at home; Lukens admits that their protocol is "very tedious." Yet behavioral intervention remains the only treatment with "well-documented empirical support" for pediatric feeding disorders, according to an evidence review of forty-eight single-case research studies spanning forty years, published in 2010 in *Clinical Child and Family Psychology Review.*

Around the time I give up on breast-feeding, our pediatrician's office connects us to Lynne Cross Menard, a speech-language pathologist at MidHudson Regional Hospital in Poughkeepsie, New York. Menard draws on the behavioral model, but whether by training or by personality, she is much more laid-back. She shows us how to tap on Violet's cheeks with our fingers or a small teething toy and then make our way over to her lips, encouraging her to gum the toy as long as she can tolerate it. Gumming toys is one of the first ways that most babies begin to use their mouths beyond milk feeding. But Violet never voluntarily brings objects to her mouth, so we have to try to do it for her. Sometimes the game lasts only seconds before Violet begins to scream and gag. Whenever she does, we stop. The idea of "replacing" bites, even when the bite is a toy, doesn't sit well with any of us. "Our goal is to give Violet positive associations with her mouth," Menard tells us.

But real progress is impossible because Violet's NG tube is worsening her aversion daily. It's hard to know what living with a tube in her nose twenty-four hours a day feels like; some doctors tell me the babies don't even feel it, while some adults who've had one report a constant burning sensation in the back of the throat.

Given how often Violet grabs at hers, we assume some degree of chronic discomfort. And we have to replace the tube every two weeks. So every other Friday, I pin her down and sing "You Are My Sunshine" while Dan threads a new tube down to her stomach. When Violet screams so hard that her throat closes, we wait until she breathes again. When she chokes and sputters until the tube comes out of her mouth, we start all over, hoping that Dan doesn't twist it into her lung by mistake. Even trained hospital nurses misplace feeding tubes as many as eight thousand times a year, according to estimates by the American Society for Parenteral and Enteral Nutrition. These mistakes can cause serious complications or even death. Nobody tracks how frequently parents do the same thing at home. Dan can steel himself to keep going in these moments only by pretending that Violet isn't his daughter at all.

A few months after we start therapy with Menard, she gently suggests that we put the bottle away. Even with breast-feeding behind us, there seems to be something about the smell or sight of milk (now, mostly formula) that is making everything worse. Then, when Violet is five months old, a surgeon cuts a hole in the side of her abdomen and implants a permanent gastric feeding tube directly into her stomach wall. This will be easier to live with; nothing taped to her face, no more torturous tube replace-ments, just a small plastic button next to her navel. Menard knows it will be our best shot at healing the aversion, but we see it only as another failure. "We are nowhere," Dan tells me the night we decide to do the surgery. We can now plug Violet in for food as if we were charging an iPhone. It is devastating. And also a relief.

Somewhere down the Google rabbit hole of those late-night pump-and-research sessions, I discover not only therapists who offer a more moderate take on the behavioral model, but also programs that reject it outright. These therapists believe that all children have that instinct to eat, as well as an innate ability to effectively self-regulate their intake. "This is a scientific conversation, but it's also a deeply philosophical one," says Suzanne Evans Morris, a speech-language pathologist and founder of the New Visions education and therapy program in Faber, Virginia, when I call her to discuss Violet's case. "Does an aversion or a complicated medical history erase a child's internal motivation to eat? Or can we help them rediscover it? I believe there is a tremendous amount of wisdom in these little kids and that they will transition to eating in their own time if we give them the right kind of support."

Morris is a kind of guru in the speech-pathology world; along with the pediatric occupational therapist Marsha Dunn Klein, she wrote *Pre-Feeding Skills: A Comprehensive Resource for Mealtime Development*, the 798-page bible of speech-language pathologists. She also helped write the profession's standard skills checklist, which Menard uses to evaluate Violet's oral aversion, and pioneered a "child-centered" approach to feeding therapy. Rather than systematically rewarding and replacing bites, Morris helps parents to read their child's cues and offer food only when clearly invited to do so. To encourage children to issue such invitations, Morris turns food into play. She might race crackers on top of a child's toy car, or for one newly and proudly potty-trained client, read picture books about the digestive system; the child's enthusiasm for eating greatly increased when he realized it would lead to poop. Making mealtimes fun in such ways may sound similar to the behavioral model, but Morris only uses such strategies to create a pleasant atmosphere and reduce a child's anxiety around food; she never sets goals for how many bites or ounces a child must consume to receive their reward. That's because her primary mission is to help children reconnect with their ability to feel their own hunger and their innate desire to eat.

The child-centered model is not without rules or structure. Disciples of this approach teach the "Division of Responsibility," a concept developed in the 1980s by Ellyn Satter, a registered dietitian and family therapist. Medical professionals—and indeed, our entire culture—have long viewed over- and under-eating as separate issues, the former rooted in a lack of willpower and the latter in a self-destructive need for control. But Satter sees them as flip sides of the same coin, and fixable through the same methodology. "In both cases, these kids are reacting to distortions in the feeding relationship with their caregivers," she tells me. To restore balance, Satter says, a parent need only take responsibility for deciding what kinds of foods to offer, and, as children graduate from nursing to solid foods, when and where meals take place. She leaves children in charge of how much food to take, what order to eat it in, and even whether they eat at all. Satter and Morris both believe that this framework allows children to experience hunger—because parents ensure that enough time passes between meals for a child to work up an appetite—and satiety, because meals are consumed at tables, not in front of TVs or on the go, when it's much harder to register whether you're getting full. But it also keeps reinforcing a child's sense of autonomy around food. Because she has the power to refuse a bite, she will be less intimidated by the prospect of taking that bite, which lets her pay more attention to whether she actually wants to eat than to the power struggle playing out around her. Satter, Morris, and many other therapists argue that the preservation of that ability to self-regulate is at the crux of solving both childhood obesity and pediatric feeding disorders.

Therapists who practice the child-centered approach have published numerous books and case studies, but have conducted no controlled clinical trials, in part because the method resists the standardization you need to design a study. The lack of published empirical data raises questions with behaviorists. "I think most of the children we work with have already tried a child-led approach and it's failed," Lukens says. But Morris lists many clients who come to her after traumatic experiences in behavioral programs. And as I get deeper into this research, I meet many formally trained behavioral therapists who have crossed over to the child-centered model.

One such person is Jennifer Berry, an occupational therapist and the founder of the Spectrum Pediatrics Tube Weaning Program, based in Alexandria, Virginia. "I realized that teaching a child to eat when their body is telling them not to is not only counterproductive, it's dangerous," she tells me. Berry resists the term "aversion" altogether because it implies a dysfunctional behavior. "It's not a problem," she says. "It's an adaptive skill to know when eating isn't safe." Studies suggest that children may eat less healthfully when parents exert too much control over the process. For example, in a 2006 trial published in the journal *Appetite*, children instructed to "finish their soup" complied grudgingly but still ate less than children who were unpressured. Research on eating-disorder patients also suggests that highly pressured mealtimes early in life might play a role in the development of those conditions.

But the behavioral model gets tube-dependent children to eat. Doctors treating such kids look at numbers—pounds, calories, milliliters—and send parents off on a panicked quest to achieve often unattainable goals by any means necessary. When the stakes are that high, worrying about the emotional consequences of the approach may feel like a luxury or, at best, an afterthought. Meals have already become fraught; comfort has been erased from food. "The question of when children start to enjoy food for its own sake remains a mystery to all of us in the feeding world," Lukens says. "Never once do I say to a parent, 'By the end of the four weeks, I think she'll *love* food.'"

All I want is for Violet to love food. It's not just that I think this is what babies are supposed to do. It's also that I love food, both for itself and for the community it inspires. This is true even though I don't come from generations of foodies; my cultural roots are British and Midwestern American, neither of which offers much in the way of proud or exotic food traditions. I do love Marmite on hot buttered toast, but am less sentimental about corn dogs, horseshoe sandwiches, or Jell-O salads, perhaps because my beloved Midwestern grandmother made no secret of her own impatience with cooking. When she and my grandfather moved to an assisted-living facility with a cafeteria, I asked whether she missed preparing her own meals. "Like a hole in the head!" she replied cheerfully. My British grandmother was similarly pragmatic: cooking was a daily chore to check off her To Do list, and food a resource to be managed efficiently, not a mode of creative expression. But my mother, perhaps in response to the postwar dinners of her childhood, enjoys eating and cooking recreationally. We ate in restaurants often during my childhood because her work schedule didn't leave time to cook on weeknights. Yet we established our own food traditions all the same, going out for sushi or Indian food, and borrowing recipes for gazpacho and pasta sauce from cultures more adventurous than our own.

And so, from my teen years on, I have loved trying new restaurants and hosting dinner parties. I went through a bread-baking phase. I've taken cooking lessons in Thailand and Italy, and can make a serviceable red curry and really stellar pasta sauce, as well as a beef stew that once inspired Dan to compose a song about it. The summer I was pregnant with Violet, we planted blackberry and raspberry bushes because I imagined our daughter chubby, barefoot, and purple-mouthed in our garden every August, eating her fill. I want to bake birthday cakes and cook Christmas dinners that will become indelible memories of Violet's childhood. It terrifies me to think that she might grow up not caring about any of that, not connecting through food. Even though we've switched to a permanent feeding tube, I want to believe we can unlock her internal motivation to eat.

Still, whenever Dan and I discuss the idea, we get stuck. How can you trust a child to eat enough after she has shown you that she is willing to choose starvation—even if that was what Morris and Berry would call "a good choice at the time"? For that matter, given how badly I had misread her cues in that first month, how can I trust myself to understand the choices she's making now?

In Palatine, Illinois, Joey giggles as he paints chocolate pudding onto a glass sliding door with Heidi Liefer Moreland, a speech-language pathologist with Spectrum Pediatrics. We are in a Chicago-area rental apartment that the team is using for Joey's feeding tube "wean," as they call the transition to normal eating. "Kids shouldn't learn to eat in a clinic," Berry tells me. "They should learn to eat at the family table, so that's where we work." Or in the case of Joey, who is two and has depended on his feeding tube ever since his premature birth, the table is near the spot where he and Heidi are working. There is nothing that resembles force-feeding; I never see either therapist even hold a spoon. Mostly, Berry and Moreland are here to hang out with Joey and his family while everybody waits for him to realize that he hasn't been tube-fed more than a few ounces per day in almost a week—and that eating is the only way to erase his nagging hunger.

This is where Berry's approach splits off from the traditional child-led model and becomes radical. Calories from the feeding tube are cut significantly over a five-day period so a tube-fed infant or child begins the wean on around 50 percent of his normal daily caloric intake and 80 percent of his optimal fluid needs. Over a ten-day period, Berry and Moreland are on call around the clock, giving support and coaching as the child, they say, rediscovers the drive to eat. After the initial wean, therapy continues as needed for six months. By then, Berry reports that 95 percent of her patients are eating all of their daily calories by mouth.

Many behavioral programs also incorporate a modified version of this "appetite manipulation": Ivy's tube feeds are gradually reduced by 60 percent over the course of her month at the Children's Hospital of Philadelphia, and Lukens credits her hunger, in part, for their success. But only a few mainstream programs (including one at Seattle Children's Hospital) have adopted a child-centered, hunger-based approach to tube weaning. Most consider it too risky because of the potential for weight loss. Berry's program requires a doctor to set safety parameters for each child and to sign off on the amount of weight loss that can be tolerated.

Later that night, Joey sits listlessly on the couch with his mother, Angela Reid. A bag of Terra Chips and an Elmo sippy cup rest on the coffee table in front of them; every so often, Joey picks up one or the other and moans.

"Is this normal?" Reid asks Berry. "He's never like this. What's going on?"

Berry sits against the far wall, assessing the situation. "I think he's very, very hungry," she says. "We have to decide whether he's going to find this motivating, or does he need a little support?" After a brief discussion, Reid feeds Joey 30 milliliters of Pedialyte through his gastric tube. Within minutes, he bounces up, all smiles: "Elmo!" By the time we leave, he has nibbled three chips.

"Usually that's just enough to take the edge off that horrible new feeling of hunger," Berry explains to me later. I ask her how Joey's palpable frustration is any different from Ivy's slack-jawed compliance with her feeder. "Because we're following his cues," she says. "We didn't let him suffer. But we didn't force him to eat before he was ready, either." The next day, Joey eats more than twenty chips. A few days later, he discovers a love of fried chicken. The family goes home a day early; over the next month, they use his feeding tube for occasional supplementation, but Joey continues to eat well and gains four ounces.

We offer Violet her first bite of banana when she is five months old, about a week after her permanent feeding tube is placed. She sits on Dan's lap at breakfast and seems fascinated as she watches us eat. I hold my breath as I offer the spoonful: Why would this time be any different? But Violet takes a taste. Solid foods intrigue her. They offer new smells and textures. They don't trigger her aversion in quite the same way.

We begin to put Violet in her high chair every time we sit down to a meal, trying to catch on quickly to the slight shake of a head that means no before it escalates to gagging and crying. We don't always succeed. At times, we offer books or toys to reward Violet, or at least to maintain her interest at the table. She doesn't eat much of anything, and the food that does pass her lips is usually gagged on or spat out. Still, the situation feels close to normal, the way our friends sit at the table with their babies and all those cups of Cheerios and spoonfuls of purées.

Then, a few weeks later, Violet undergoes her second open-heart operation. Over the next three months, we spend fifty days in the hospital as she fights off various complications. Eating is forgotten. Often, she's too sick to keep down her tube feeds, or so frail that she can only be fed intravenously. I feel all the progress we've made with those tiny bites of banana evaporating. But one day, Violet takes a previously rejected sippy cup and drinks an ounce of water. She is hooked up to oxygen to keep her breathing, and a nurse explains that breathing on a nasal cannula can feel like driving down a highway with your head out the window. So Violet is thirsty for the very first time—and somehow, she knows that she can use her mouth to make herself feel better. Our feeding therapist, Lynne Cross Menard, texts me: "We're still in the game!!" We now have a bit of proof that Violet's internal drive is still there, somewhere.

After Violet recovers, we go home and spend the summer watching for any sign that she is ready to try eating for real. We begin working with a pediatric dietitian named Margaret Ruzzi, who instructs us to offer table foods before every tube feed. By her first birthday, Violet is taking a few bites of food at every meal. She loves flavor—chicken tikka masala, pad thai, blueberries. She will thoughtfully chew lemon wedges. One afternoon, she snuggles on my lap and happily gums an apple core. It's the first time I've held her to eat since we stopped breast-feeding. Food once again means comfort. The aversion is gone.

As Violet reaches the age when most children transition away from breast milk or formula, Ruzzi encourages us to change what we put in her feeding tube accordingly. This is a bold move. Many dietitians and doctors believe that tube-fed children should continue on formula more or less indefinitely. During our hospital stay, one gastroenterologist mapped out how she expected Violet to transition from infant formula to formulas designed for toddlers, then older children and adults. When I asked about whether we could ever blend up real food for her, she sighed. "It's not a good idea. It really clogs up the feeding equipment." There was no discussion of whether Violet might benefit.

But Ruzzi and Menard are big proponents of putting Violet on a "blended diet." Not surprisingly, so are child-centered advocates like Morris, who self-published the *Homemade Blended Formula Handbook* in order to help parents understand how to begin this process. We start by testing out small samples of food—an ounce of applesauce or yogurt blended into Violet's usual infant formula.

When that seems to go okay, we step it up. Soon I have purchased a red Vitamix at a medical discount, and am blending ground beef, quinoa, blueberries, and kale into almond or rice milk. I spend hours prepping and freezing ingredients every weekend so we can blend them up each morning. The gastroenterologist was right about one thing: blended food does clog the feeding pump. We retire it and switch to pumping each feed into Violet's gastric tube by hand using giant syringes, which I fill each morning so we can feed Violet throughout the day. It's messier. There are blended-food stains on all our clothes, the floor, the ceiling. And these are not artisanal hipster smoothies; blended meals look and smell like vomit. But interestingly, Violet's vomiting, which mysteriously began to slow down a few weeks before we began the transition, stops entirely once we introduce real foods. The whole process feels closer to what we're supposed to be doing. I love making the blends, deciding which fruits or vegetables to try. I don't miss the whirs and beeps and error messages of the pump. "I'm finally feeding my baby," I think.

This is a hard realization. My milk supply was decimated by the combined trauma of Violet's near-death experience and the abrupt end of our breast-feeding relationship, but I still agonized over the decision to stop pumping and switch her over to formula entirely. Really, there was no choice. Some mothers are able to exclusively pump for months; I stopped when the return on a forty-five-minute session dropped from two ounces, to one ounce, to less than a teaspoon. I found myself chasing even the tiny bits of breast milk that splattered onto the pump's plastic flanges, tilting the funnels this way and that, trying to direct each wayward drop down into my storage vials so it didn't go to waste. I did this until there were no more drops to chase. "It's all about nipple stimulation," the lactation consultants told me. But my body seemed to know the difference between the pump and a baby. It seemed to understand how few milliliters Violet was willing to drink by mouth, and slowed production down in solidarity. So we became a formula household.

And again, I did the research. I learned that breast-feeding is associated with some small gains in cognitive development, and perhaps a reduced risk for childhood obesity, ear infections, and other health problems. But most of the available studies were conducted in such a way that it's hard to parse out which benefits come specifically from breast milk, and which come because families that can make breast-feeding work are more likely to have so many other advantages, such as education and financial security. We couldn't make breast-feeding work, but we are fortunate to have such resources. I was determined to let go of the "liquid gold" rhetoric and be okay with formula.

Yet, as soon as I read about blended diets, I am sold. To Dan, to Ruzzi, to Violet's doctors, I say it's because my research shows that a blended diet will help prevent Violet's vomiting and support our goal of oral eating by priming her digestive tract with the same kinds of foods as we're offering at the table. But I also want to feed my baby real food. Even while I resent the implication that formula is not food—that it's somehow less worthy than breast milk—I still want to feed my daughter something more than cans of powder. I cry when other parents post proud, messy photos on Facebook of their babies in high chairs, happily covered in spaghetti or carrots. I rage privately when friends wax poetic about baking first-birthday cakes sweetened only by apples and maple syrup. "How can they deprive their kid the pleasure of real sugar," I think, "when they're lucky enough to have one who wants to eat it?"

Yet I am the same type of mother. If things had gone differently, I would have been making pink frosting out of beets and agave nectar, right alongside them. Transitioning to the blended diet is the "tube mom" equivalent of making my own baby food. It is time-consuming, labor-intensive, and expensive. There is no hard evidence that it's really better for Violet. But I've lived too long with shame and anxiety about feeding my child nothing but powder. I feel constant guilt over food: every healthy food she won't eat, and then, as we start to see the small signs of progress, whether her newfound affinity for Cheez-Its represents an even bigger failure than our inability to get her to eat in the first place. The world of pediatric feeding disorders is, in so many ways, a funhouse-mirror version of American food culture.

So I prep and freeze and blend, and Violet continues to receive all of her calories through her feeding tube. Because she never gets hungry, eating by mouth is purely recreational. Sometimes it feels as if she's baiting us: she packs spoonfuls into her cheeks and then—just as we think, "Here is true eating, at last!"—she spits it all out. I begin to understand why food refusal is so often classified as a behavior problem. We reprimand the spitting as if our fifteen-month-old is displaying bad table manners, but then I worry we've introduced a new stressor to family mealtimes, when we're supposed to be cultivating a relaxed, joyful atmosphere. I describe the constant spitting to Morris, who suggests that I reframe it as a critical part of Violet's learning curve. "Spitting helps Violet know she can get the food out," she explains. "That makes it safe to experiment with taking another bite." So we let Violet spit. And start to think about what might happen if we starve her a little.

It feels counterintuitive and maybe even self-indulgent. I wonder if we're pushing too hard because the feeding tube—and all of that blending, syringing, and cleaning—is driving us insane. But Violet isn't eating because she doesn't know she needs to eat. And so with Ruzzi and Menard's support, we decide to drop her tube calories by 20 percent for two weeks.

Our plan is to start by cutting a "morning snack" tube meal, which Violet normally receives at nine o'clock; instead, she would have six hours between breakfast and lunch to experience hunger for the first time in over a year. During the first two weeks, not much happens. Violet cries more, particularly around eleven a.m., when presumably her stomach feels empty. But she has no idea what to do about it. We hold her a lot. She doesn't eat more. At our next checkup with Ruzzi, she has lost ten ounces but also grown an inch. "She used her calories for growing," says Ruzzi. "That says something. Let's keep going."

So we cut Violet's tube feedings by 40 percent, and over the next two weeks, Violet begins to eat. A tiny wheel of Babybel cheese. A pouch of applesauce. At our next check-in, she has gained five ounces. We cut back more on the tube calories. By November, we estimate that Violet is eating between 150 and 200 calories by mouth per day. In December, I inject her final tube meal.

Not long before that last tube-feed, Dan, Violet, and I go to a diner for brunch. It's the first time in over a year that we've gone anywhere without packing the pump, syringes, and tube. We order egg sandwiches and, from the kids' menu, a grilled cheese, which Dan carefully cuts up into postage-stamp-sized bites. All around us, other families are tucking into their Sunday pancakes, chatting and clinking forks. Violet scribbles with crayons on her place mat, throws my french fries to the floor, giggles at her dad's funny faces. And then she eats everything but the crusts.

July 2015. Violet is almost two, and she eats now. This is still new, tenuous, and astonishing to me. It's nearly eight o'clock in the morning; I should be unloading the dishwasher, finding the sunscreen, and doing everything else that parents do to get their children out the door on busy weekdays. Instead, I linger at the table, refilling her small ceramic bowl with more Cheerios so I can watch her grasp the spoon again with her now quite chubby hand and bring another sloppy bite to her mouth.

When she finishes chewing, Violet says, "More chocolate milk?," only with her toddler accent, it sounds more like "Moh choc-la melk?" And then we practice saying "More, please!"—"Moh choc-la melk peez!" with a satisfied head nod—as I hand over her blue sippy cup. "Boo cup!" she notes approvingly as she grabs it with both hands.

Violet eats two bowls of Cheerios and three bites of banana and drinks six ounces of chocolate milk before we leave the table. She's finally in the midst of a long-hoped-for growth spurt, and I'm watching sundresses that hung below her knees in May become miniskirts as August approaches. And while she likes many foods that mothers like me are supposed to feed our kids—kale, quinoa, berries—I know the real secret to Violet's growth is that chocolate milk, full of fat, sugar, and 210 beautiful calories per cup. We go through a half gallon each week.

But back when we were first desperately trying to tempt Violet to drink out of a bottle, cup, or straw, chocolate milk was something I refused to serve. Even though it was obvious that she hated the taste of plain milk. Even though I understood why: after all, she had vomited up milky-tasting formula at least four times a day for the better part of a year on her feeding tube. Pavlov himself couldn't have conditioned her more classically to dislike unflavored dairy. So I offered almond milk, coconut milk, and lots and lots of water. Every week in the grocery store, I lingered in front of the chocolate milk, available from the same local, hormone-free dairy that provided our plain whole milk. Sometimes I put a carton in my shopping cart, then back on the shelf. I wasn't entirely sure she'd like it, anyway; at the time, Violet was refusing all sorts of foods that conventional wisdom assumes babies and children love, including yogurt, mac and cheese, and anything puréed. But a small part of me was sure she would like it. And that if she did, this would be bad.

When I was growing up, chocolate milk was the only kind I would drink. My mother also can't stand the taste of plain milk and used to get in trouble for not drinking it at her British primary school, so she supported my need for several dollops of Hershey's Syrup stirred into every cup. And this was the 1980s, when sweetened milks were newly trendy and not yet reviled. Fat was our dietary enemy, not carbohydrates, so sweetened skim milk was widely embraced, a staple on school lunch trays across the nation. But more recently, chocolate milk has become symbolic of everything that's gone wrong with childhood nutrition. School food advocates talk about how it contains as much sugar as soda, and when Michelle Obama launched her "Let's Move!" campaign to combat childhood obesity in 2010, getting flavored milk off school lunch menus was a key objective.

Obama and others have argued that giving kids ready access to overly sweetened foods such as chocolate milk conditions them to prefer hyper-sweet foods and be pickier about other flavors. On a personal level, I knew this could be true; plain milk never made me gag, as it did my mom. I just liked the sweetened kind better, because I liked sweet everything better back then. As a little kid, I lived on pink Yoplait strawberry yogurt, Smucker's strawberry jelly sandwiches on white bread, Chips Ahoy chocolate chip cookies, and ziti with Ragú tomato sauce. And there is some research to back up my anecdotal experience. Children in a 2004 study who were given a sweetened orange drink for their midmorning snack eight days in a row not only liked it better than kids who weren't continually exposed to the flavor, but also opted to drink more, even after the eight-day trial period. Other research shows that, in lab tests, children who regularly eat sweet foods show a stronger preference for sugary flavors than do children who aren't exposed as often. The brains of overweight children also seem to light up more in response to sweet flavors than do the brains of thin kids. And the link between sugar consumption and metabolic disorders such as diabetes is well established.

Based on all this, I thought serving chocolate milk to kids was at best kind of lazy and at worst potentially dangerous. Only in hindsight can I see how absurd I was being. None of that research applied to us. My own early years as a garden-variety picky eater were a world apart from what we were dealing with. Violet wouldn't eat. She was barely on the growth chart. Her hummingbird heart burned more calories in a day than we could pump into her tiny body. And yet I worried that if I fed her chocolate milk, she would like it too much. I thought that on top of teaching her to eat, I also had to teach her to eat perfectly. Otherwise, I imagined, she would never develop a taste for plain milk and other low-sugar foods, and we would pave the way from a feeding disorder directly to childhood obesity and diabetes. Making the easy choice now could mean that one day, she might get fat.

Food is supposed to sustain and nurture us. Eating well, any doctor will tell you, is the most important thing you can do to take care of yourself. Feeding well, any human will tell you, is the most important job a mother has, especially in the first months of her child's life. But right now, in America, we no longer think of food as sustenance or nourishment. For many of us, food feels dangerous. We fear it. We regret it. And we categorize everything we eat as good or bad, with the "bad" list always growing longer. No meat, no dairy, no gluten—and, goodness, no sugar. Everything has too much sugar, salt, fat; too many calories, processed ingredients, toxins. As a result, we are all too much, our bodies taking up too much space in our clothes and in the world. Food has become a heavy issue, loaded with metaphorical meaning and the physical weight of our obesity crisis. And for parents, food is a double burden, because we must feed our children even while most of us are still struggling with how to feed ourselves. When the feeding tube first went in, I thought the hardest part of teaching Violet to eat again would be persuading her to open her mouth. Actually, the hardest part was letting go of my own expectations and judgments about what food should look like—so I could just let her eat.

Two

Chasing Clean

I meet Christy Harrison in a crowded Brooklyn bar on the first night of summer in 2017. She's an extremely pretty brunette with chic, angled bangs and a great navy-striped dress. There's also something just a little bit guarded about her. As we talk over pork belly sandwiches, I decide it might be from knowing she appears to fit seamlessly into one mold—stylish, thin, healthy-food-oriented—when, in fact, she's been working hard for many years now to build a different kind of reputation. "I was a food writer," Christy says, "who was really struggling with food."

Christy was on staff at *Plenty*, a hip indie magazine that ran for a few years in the mid-2000s, and later at the iconic *Gourmet* magazine in the final two years before it folded in 2009. She wrote about organic farming, biodynamic wine, and the evils of processed foods. "Oh, and I definitely helped fan the flames of the gluten-free craze," Christy tells me. She's not proud of this. Because in between all those fancy press lunches, chef interviews, and junkets on organic farms, Christy was spiraling. What began as a diet in college turned into a full-blown eating disorder by her early twenties. "My symptoms never fit neatly into one eating-disorder diagnosis," she explains. "I would move between restricting, bingeing, and over-exercising, over and over, but it was always sort of nebulous." Christy can only see the patterns of her disorder now, in retrospect. At the time, she thought she was just "obsessed with food," and really, no different from anyone else around her.

Christy and I didn't know each other during those years, but her story is familiar. I know the world she inhabited, because I was at its periphery, as a junior health editor at the now defunct *Organic Style* magazine, and later, as the kind of freelance writer whom women's magazines call when they need a piece on which types of produce are the most important to buy organic, or why you should learn to cook quinoa. On one freelance assignment in 2007, I found myself sitting at the bar of the famed farmhouse-chic restaurant Blue Hill at Stone Barns, in Pocantico Hills, New York. It was noon on a Wednesday and the restaurant was closed, but I was eating the most delicious salad of my life, hand-prepared by the celebrity chef Dan Barber with greens picked that morning from his greenhouse and wild mushrooms foraged from the forests around the farm. I was definitely not then (nor am I now) the type of person whom Dan Barber makes lunch for on Wednesdays. But I was there doing research for a book with someone who was—a woman who, at the time, was a little bit famous for her philanthropy and her extreme vegan politics.

On the drive up to Stone Barns, this well-known vegan listed out all her favorite vegetables and told me how much she loved avocado, despite its high fat content. "My son and I share half an avocado every morning for breakfast," she said. "It's worth the extra calories because it makes your skin glow." After we ate our salads, the waiter brought out a plate of fresh-baked chocolate chip cookies for dessert. "Are those made with white sugar?" she asked. "I think we're all done." I had only eaten six mushrooms and three handfuls of micro-greens and I didn't know about eating avocado for breakfast. The cookies, the waiter explained, were made with fair trade dark chocolate and butter from a cow named Tallulah. The vegan demurred again, this time citing her sympathy for Tallulah. Her assistant and the restaurant's publicist followed suit. I took the cookie. Even the waiter seemed surprised.

Back then, I focused on eco-food issues because I thought that niche would help me avoid writing the kind of weight-loss stories that women's magazines are so notorious for running every month. I wrote those, too—when you're a broke freelance writer, there's not much you won't write about—but always with a degree of existential crisis and bargaining: I would write them for women's magazines, but not for teen magazines. I would write one about friends pairing up to lose weight together, because that felt at least vaguely empowering. I would write one about the relationship between blood sugar and white sugar because that seemed scientific and important. But eventually, I stopped wanting to write any such stories; I hated trying to find definitive answers about weight loss when the science was constantly shifting. And I hated telling women how to make their bodies smaller when I didn't really believe they should. And so I embraced the eco-food movement, because—on the surface, at least—it wasn't about calorie counting or crash diets. It was okay to eat the cookie, I thought, as long as it was made with fair trade chocolate and local butter and eggs. But as I learned that day at Stone Barns, it still wasn't okay to eat the cookie, no matter how sustainably sourced the ingredient list was.

This is because the eco-food movement, also known as the eco-gastronomy or alternative-food movement, was busy embracing the war on obesity, joining the front lines of the fight. And food became something to categorize—whole or processed, real or fake, clean or dirty—and to fear. Pretty soon almost every food and health writer I knew was dropping gluten or white sugar from her diet, then bringing it back, then dropping something else. Now that trend has gone mainstream; even my eighty-eight-year-old grandmother knows what gluten is and why half her family isn't eating it on any given day. And that's because the early to mid-2000s represented a kind of cultural awakening around food, driven by the success of movies like *Super Size Me* and books like *Fast Food Nation* and *The Omnivore's Dilemma*. To be a food journalist, especially in New York City, at the glossy magazines that Christy wrote for, meant that you had be up on who made the best artisanal iced coffee and designer cupcakes, but also on the latest research showing why we're not really meant to digest dairy, or the differing omega-3 quantities of grass-fed and grain-fed beef. It all felt really important, as if we were on the verge of discovering a profound connection between nutrition, health, and the environment that nobody had ever seen quite that way before.

Connecting all her food anxieties to a larger movement made Christy feel "like I wasn't just this vain, selfish person trying to lose weight." Suddenly, her obsession with food had a larger, nobler purpose. "I thought, 'I can help change the landscape and make everyone healthier,'" she said. "There was some desire to be communal about it, a kind of social justice drive." Giving up meat and gluten wasn't dieting—it was a hard and virtuous life choice that would transform your health and boost your green street cred at the same time. But in the hands of those celebrity chefs, food activists, actresses turned health gurus, and the media who love them, "eating clean" involves trading one set of food rules for another. "Now I think about how much I was struggling and so were all these other food writers I know," Christy says. "We kept thinking we were finding answers. But really, we were participating in this mass marketing of disordered eating."

It took several more years before Christy came to that realization and began to think critically about the alternative-food movement and the "clean eating" craze that it helped to fuel. As a food writer, she continued to cycle between the extremes of her disorder. She dated another food writer, and sometimes it seemed as if being a part of that world was helping her break through some of her most restrictive rules. Because a big part of the job is going out to eat exotic, lavish meals, at some point a food writer who doesn't eat just starts to look weird. "But I'd compensate by restricting even more when I was on my own," she notes. And when *Gourmet* folded in 2009, Christy was still deeply passionate about helping to spread the gospel of whole foods, so she decided to get her master's in public health and become a dietitian.

At first, studying nutrition all day only further entrenched Christy's eating habits. She wanted to do well in school, and it seemed as if eating perfectly was part of being a model student. But one day in class, everyone was told to partner up and take each other's body measurements. They got on scales, they wrapped tape measures around their waists, hips, and necks, and they used calipers to measure the fat on the undersides of their arms. Obsessively tracking body size in this way is the kind of thing that people with eating disorders do all the time; so much so that putting away the scale and the tape measure is often an essential first step in any treatment program. Yet it's also a core part of many nutritional studies programs, most of which are built on the belief that body weight is our best tool for assessing physical health.

Although Christy was far from recovered, she had made enough progress to know how dangerous it was for her to obsess over numbers like these. But refusing to participate didn't feel like an option. So she wrote down all her measurements. Then she looked in the textbook to see how her numbers compared to the "ideal" weight for someone of her height and build. She was, as she puts it, "more than a few pounds" over the textbook limit. "At first, I panicked, like, 'Okay, I've got to really double down and eat even better,'" Christy recalls. "Then I realized: the only time I'd been that small was when I was in the most intense restriction period of my eating disorder." Dietitians are trained to think in terms of numbers— calorie counts, body fat percentages, body mass indexes—and refer to textbook ideals for each metric. But for Christy, the textbook math didn't apply. And though she's deliberate about not sharing exact numbers (precisely because of how triggering they can be) it's worth noting that Christy is not a large person. "I've never experienced discrimination based on my weight," she says. "So if someone with an enormous amount of thin privilege is still considered significantly above this 'ideal,' I can only imagine how much more horrible and shaming this exercise is for people who live in larger bodies." She was done. "That's when I decided to just throw out that whole model of thinking about food and weight."

The calorie-counting, body-fat-analyzing approach to food and weight taught by most dietetics programs differs from the approach of the alternative-food movement and in many ways conflicts with it. In the alternative-food camp, it's often argued that as long as you're eating the right foods (organic, locally grown fruits and vegetables, gluten-free grains, and maybe some sustainably sourced almond milk or raw cashew cheese), you can stop worrying about portion size. Or at least you can subscribe to metrics like Pollan's "Eat food. Not too much. Mostly plants," which, ironically, can feel even more difficult to adhere to because they're so deliberately vague. (How much is "too much"? Do carbs count as plants?) When I reported nutrition stories in the early 2000s, conventional dietitians were often rattled by the organic foodie approach, and kept pointing out the surprisingly high calorie count of foods like olive oil, quinoa, and almond butter. But over the past decade, the two camps have slowly merged. The Academy of Nutrition and Dietetics spokespeople whom I interviewed for *Redbook* and *Runner's World* became more fluent in the language of clean eating. They began talking less about how to choose chicken breasts the size of a deck of cards, and more about the nutrient profiles of various nut butters and how to cook whole grains. Meanwhile, the newer generation of alternative foodies has become much more interested in things like protein and carbohydrate grams, and how many of each you should or shouldn't eat per day. And one core philosophy has always united both approaches: That we are what we eat. That food is medicine. And that there is no problem—be it constipation, migraines, infertility, or cancer—that cannot be solved, or at least vastly ameliorated, by changing your diet.

There is a kernel of truth in almost every one of these notions. The industrialization of agriculture and food production has led to an overabundance of high-calorie foods, many of which are marketed with a veneer of health because they're "low carb" or "gluten-free," while also being largely nutrient-free. Producing these foods causes significant environmental damage and involves the use of chemicals and ingredients that aren't great for our health. And some folks have a tougher time with this onslaught than others. About 1 in 140 Americans have celiac disease, for example, which is a severe autoimmune disorder in which eating gluten triggers the immune system to attack the small intestine, causing permanent damage and malnutrition. And 65 percent of the world has some degree of lactose intolerance, according to National Institutes of Health data. Many people in that group likely hail from cultures that don't drink milk past infancy, and their bodies have a hard time digesting its proteins.

The problems begin when we consider the corollaries to statements like "You are what you eat." If that's true, then eating "bad" foods (Big Macs, Slushies, anything made with white flour or sugar) makes you a bad person. Or at least an uninformed, undisciplined one. Organic farmers and food activists may have originally banded together to take on huge corporations within the agricultural-industrial complex. But infusing their arguments with messages about health has led to the rise of a wellness-industrial complex, in which nutritionists, personal trainers, cookbook authors, and other "alternative-health experts" target us for our individual choices. They aren't fighting evil corporations like Walmart or Amazon anymore. They're hoping those evil corporations will stock their products. Alice Waters, the founder of Berkeley's iconic Chez Panisse, wrote on Twitter to Amazon's CEO, Jeff Bezos, "with hope-fulness" after the online giant announced it was buying Whole Foods in the summer of 2017. "You have an unprecedented oppor-tunity to change our food system overnight," she tweeted. "And we are all here to help you do it!"

Alternative food and wellness are big business now. The Amazon–Whole Foods deal was worth $13.7 billion. Sales of old-school diet staples like Lean Cuisine meals may have dropped by $100 million between 2014 and 2015, but expensive, largely organic meal-delivery services like Blue Apron generated close to $1.5 billion in sales in 2016. The Global Wellness Institute, a nonprofit based in Miami, Florida, which conducts industry research, calcu-lates that the worldwide "wellness economy" is now worth $3.7 tril-lion. They attribute $999 billion of that to beauty and anti-aging products, and another $648 billion to "healthy eating, nutrition and weight loss." And the marketing around these products and services is just as powerful as any fast-food ad campaign.

We are now so certain that every aspect of our health can be improved through diet, we can only blame ourselves when those diets fail. When cutting out gluten doesn't work, we move on to dairy, then soy. When we still don't feel better, we start reading about the evils of nightshade vegetables or peanuts. Still feel bloated, or tired, or lacking in energy—all impossible-to-quantify symptoms that may just reflect the unavoidable state of being mortal and not part superhero? Probably it's because you weren't careful enough about that gluten. Nutrition has become a permanently unsolvable Rubik's Cube. So we read more books, pin more blog posts, buy more products, and sign up for more classes and consultations. And we don't realize how many of the so-called experts guiding us through this new and constantly changing landscape are exactly where Christy once was—fighting their own battles with food.

Today, Christy still works as a dietitian, but she has dramatically revamped her practice. She worked in-house at two eating-disorder recovery centers in the New York area, and now coaches private clients, people who have gotten caught up in our culture-wide fixation with dieting and detoxes, and want to find a way out. She also hosts a weekly podcast, *Food Psych*, in which she interviews other dietitians, therapists, and people in the wellness industry who are trying to fight back against what Christy alternately refers to as "diet culture" or "the thin ideal." *Food Psych* is well ranked on iTunes' list of the Top 100 Health Podcasts, but when you look at the rest of the list—with names like *Half Size Me*, *Livin' La Vida Low-Carb*, and *Vegan Body Revolution*—you realize just how hard this thing is that Christy is trying to do.

On the podcast and throughout our conversations, Christy uses the term "orthorexia" to describe the obsessive thoughts and restrictive habits that dominated much of her twenties. Orthorexia is not an official eating disorder diagnosis in the *Diagnostic and Statistical Manual of Mental Disorders*. Steven Bratman, M.D, an occupational medicine specialist now based in Fairfield, California, coined the term in 1996. "I originally invented the word as a kind of 'tease therapy' for my overly diet-obsessed patients," he writes on his website, Orthorexia.com. "Over time, however, I came to understand that the term identifies a genuine eating disorder." Bratman defines orthorexia as "an unhealthy obsession with healthy food," and says that the onset and progression of the disorder closely mimics the official diagnostic criteria for anorexia, except that a preoccupation with health and "clean" eating serves as a proxy for an obsession with weight loss. "Most people with orthorexia would say it's not about the weight," notes Emily Fonnesbeck, a dietitian in St. George, Utah. "They'll say, 'Oh, no, it's just about wanting to eat really perfectly,' whereas someone with anorexia would say it's about getting fat. But in recovery from orthorexia, you uncover that it really is all about the weight. The thin ideal is underlying in every eating disorder."

Emily knows this from her work with eating-disordered clients—and because, like Christy, she also spent years with orthorexia. Emily grew up one of six kids and she says food wasn't a big deal in her house. She ate when she was hungry and stopped when she was full. But she was also the thinnest of her siblings. "There may have been some differences in how food was approached with each of us," she says now. "And I may have subconsciously picked up on some fear of fat, and the idea that you need to manipulate food in order to not get fat." In her late teens, Emily started taking birth control pills and gained some weight. "It didn't start out as obsessive or unhealthy, but I did start to become much more concerned with eating healthier food and exercising," she recalls. "And I've always had a strong perfectionist streak." She experienced a rising tide of anxiety around food, which fueled her decision to study nutrition in college. "I think I did this for a lot of my own reasons," Emily says. "I was interested in food, but I wanted to learn about it for me, versus really wanting to help other people. It felt like being a nutrition student would help me learn to eat perfectly."

The real tipping point came at age twenty-three, when Emily had her first child. "I loved being pregnant. I've never felt better about my body in my life," she says. "But I wasn't prepared for after—for all of the physical changes, being home with this new baby. It was just totally new for me and I felt really out of control in a lot of ways." Emily found herself struggling with intense postpartum emotions, which mostly manifested in body anxiety, and then in increasingly intense exercise and eating habits. "I just needed a way to cope, and this is what I turned to," she says. But she's not sure she would have gone there without her professional training; Emily finished her dietetics internship the day before her son was born and started her first job as a dietitian nine months afterward, so her postpartum stress played out in the context of this newly launched career.

As her anxiety intensified, Emily also began to experience frequent digestive distress. And it felt like the only logical step to scrutinize her diet even more. She spent $195 on a twelve-hour certification program called LEAP (Lifestyle Eating and Performance). LEAP is marketed by Oxford Biomedical Technologies, a medical lab in Florida that claims to have performed over 4.5 million food sensitivity tests. As a certified LEAP therapist, Emily was trained to coach clients through elimination diets, and also to sell them Oxford's testing protocols. For $550, they could have a vial of blood analyzed for evidence that certain "inflammatory" foods in their diet were contributing to their migraines, irritable bowel syndrome, arthritis, or even ADHD.

Research supporting these kinds of testing protocols and diet plans is sketchy at best. In 2012, the Canadian Society of Allergy and Clinical Immunology issued a position paper generally condemning such tests, without referencing any particular companies or programs, for their lack of scientific data. The paper's lead author, Stuart Carr, an allergist at the University of Alberta, told *Outside* magazine: "[Food sensitivity] is an incredibly vague, nondescript term that has no real definition. When the companies marketing these tests use 'sensitivity' instead of 'allergy' on one of their labels, they do it because they know it doesn't really mean anything." Terms like "detox," "cleanse," and even "elimination diet" also lack official medical definitions; they're just marketing words that any self-styled wellness guru can use, whether he's selling a lab test, a protein powder, or his own "lifestyle brand" on social media. Part of the problem is that research on the inflammatory properties of any individual food is in its infancy; what scientists may be able to demonstrate in a petri dish during a perfectly controlled laboratory environment won't necessarily tell us much of anything about how the complex environment of a fully functioning human body will handle these foods when eaten in varying quantities and combinations out in the real world.

The same goes for research on specific foods like kale, watercress, and ginger, which the proponents of elimination diets love to tout for their so-called detoxifying properties. In 2015, I reported a story on the detox diet trend for *SELF* magazine and interviewed several doctors who were marketing their own books and plans. One of them, Woodson Merrell, who is an integrative medicine physician and the author of *The Detox Prescription*, pointed me to a few dozen studies that supported his program. Some were done in labs, where scientists infect human tissue samples with toxic chemicals, and then observe how compounds from plants like ginger and coriander affected the cells' ability to excrete toxins. Others were done on rats or small human groups—for example, a study funded by the National Cancer Institute, which found that eleven smokers who ate two ounces of watercress with each meal excreted higher levels of the carcinogens found in tobacco. Merrell argued that such research is a promising sign of food's ability to detoxify us. But other researchers I spoke with for the story pointed to the limits of such experimental data. We might all be better off eating more watercress, but that's about all you can say for a study done on one particular food and one small group of people.

The other big flaw with elimination diets is this: We don't need to cleanse. Our bodies are already set up to eliminate waste and detoxify us using our liver and kidneys. As blood flows through your body, it's filtered by your liver, which snatches up toxins, cholesterol, and other impurities that enter your system (often through the food we eat, but also via air, water, and consumer goods) and circulate in your bloodstream. Your liver then uses a two-phase process to clear out those unwelcome substances. First, it releases special enzymes to convert the toxic molecules into new, unstable molecules known as free radicals. Then those free radicals are bound to certain substances that essentially fast-track them over to your kidneys, so you can pee them out. (You also excrete them through stool, sweat, and even exhaling.) "The inside of your body is not dirty and it does not need cleaning," Michael Gershon, M.D., a professor of cell biology and pathology at Columbia University, told me when I asked him to explain detoxing for the readers of *SELF*. And indeed, a 2014 study review published by the British Dietetic Association concluded, "Although the detox industry is booming, there is very little clinical evidence to support the use of these diets."

For the *SELF* story, my editors convinced me to go on a detox diet myself, with the help of a chic holistic-medicine practice in Manhattan. I went in skeptical. I'd read the British Dietetic Association's review. I went in convinced that I'd see through the hype. But I left the first consultation feeling fat, miserable, and like my insides were full of sludge. The detox industry's marketing is powerful. And imbued with just enough science that it's easy to drink the Kool-Aid, or, more likely, the raw kombucha. After all, research by the Centers for Disease Control has found detectable levels of hundreds of environmental chemicals such as pesticides, flame retardants, and tobacco smoke by-products in the blood and urine of large sample groups. We don't really know how well our bodies can cope with these types of things; some, like lead, have been shown to accumulate in our bones. And that's where the wellness industry experts come in, with their promises that what you eat can be powerful enough to undo this kind of damage.

Of course, Emily underwent the LEAP testing herself. Her analysis revealed that she was "sensitive" to a laundry list of foods: dairy, potatoes, cauliflower, cherries, mangoes, apples, rye, salmon, crab, chicken, beef, lamb, pecans, walnuts, soy, peas, pinto beans, black pepper, maple, vanilla, oregano, paprika, parsley, and the amino acid tyramine. "Just typing that makes me want to scream," Emily observed when emailing me the full list. Still, when she first got the results, her response was to immediately and forever cut those foods out of her life.

This intense restriction didn't make Emily feel better. In fact, she felt worse and worse with every new food that she took out of her diet—yet that only propelled her to keep going. Food had become very black-and-white, and once a given food had been eliminated, it was bad, dirty, verboten. Emily couldn't conceive of eating it again, even as the list of acceptable foods continued to dwindle and her health deteriorated. This is a common form of magical thinking about such diets. Online forums and Facebook pages for the fans of detox programs like Whole30 and Simple Green Smoothies devote a significant amount of time and conversation to dissecting the various ways that these diets make people feel bad: tired, bloated, nauseated, headachy. But these symptoms are always framed as a necessary part of the process—as proof, in fact, that the diet is working and your body is shedding toxins or otherwise somehow purifying and morphing into your lighter, cleaner, truer self. The worse she felt, the more Emily bought into that trial-by-fire theory. "I kept thinking, 'I'm detoxing, eventually I'll turn a corner and things will feel better,'" she says. "But it wouldn't. It was never going to. One of the biggest lies my eating disorder told me was that my strength came from living by all these rules. True strength is actually saying no to things that don't serve you."

Emily reached what she considers her "rock bottom" in 2012, when she walked into her then seven-year-old's bedroom and found him furtively eating a Kit Kat in the corner. "I never really said to him that he couldn't eat a Kit Kat," she says. "But he got the message just from watching me and feeling my anxiety about food. I saw that this was becoming bigger than me, and that I had to get healthy for them." Around the same time, the family was planning a five-day vacation and Emily, whose diet was limited at that point to exactly six foods, began to panic about how she would eat while traveling. One night her husband came into the kitchen to find her grumpily packing up everything she could eat to take with them. "Emily, this is not you," he said. For the first time, Emily allowed herself to think that maybe it wasn't that she needed to eat perfectly—maybe it was that she needed to start permitting herself to eat, period.

"Now I know that psychological stress can influence digestive function just as much as what you're eating—so adding more stress through diets and food rules is absolutely going to influence digestion for the worse," Emily says. "But at the time, I didn't see my symptoms as a side effect of my restrictive eating. I saw them as proof that I needed to be even more perfect in my eating."

The cult of clean eating is dangerous enough when it targets the kind of benign health problems—headaches, constipation, fatigue, insomnia—that are perhaps just the by-products of a busy modern life. I don't mean to sound dismissive of the pain and suffering involved in any of those conditions; I developed my first aura migraine in the middle of my freshman college orientation: weird sparkly lights engulfed my vision, followed quickly by pain and vomiting. I spent much of my twenties getting migraines almost weekly. And I experimented with all the things you try when everyone from your dentist to your mother's best friend is sure they have the fix. It was a maddening and at times all-consuming ordeal, made all the worse during the months when I cut out bacon or peanut butter in the hopes it would fix everything, only to be waylaid yet again. But after that first migraine, I knew that this condition couldn't kill me. When elimination diets are marketed to people who are truly sick, with a condition that could be fatal, or that at least has the potential to devastate their quality of life, we see how dangerous this phenomenon truly is.

One day when Anna Sweeney was fifteen years old, she dunked a basketball in her friend's driveway. A few days later, her right hand started to feel weird. She couldn't feel hot or cold and couldn't get a good grip on a pencil. Anna's mom took her to a neurologist near their home in Boston, wondering whether the basketball dunk had caused nerve damage. The neurologist sent Anna for an MRI and then, when her legs started to tingle and burn, to a team of specialists at Boston Children's Hospital. Anna was admitted for five days of intravenous steroids and tests. On the fifth day, a doctor brought her into a tiny consulting room and explained they had found what are known as demyelinating lesions in Anna's brain. She was diagnosed with relapsing, remitting multiple sclerosis.

"I was pretty numb," Anna says about that day. She's now thirty-two years old and a dietitian who treats eating-disordered patients in Concord, Massachusetts. "It was this really big diagnosis, but also, after five days of steroids, I was feeling fine. And I was just a child." She was excited when all the kids at her school—including "all the hot senior boys"—signed a giant get-well card for her. Other than that, she didn't really know what having MS would mean.

But over the course of her high school career, Anna had thirteen additional "flares," during which her arms or legs would become numb or painful. And pretty quickly, she started to figure out that one thing MS meant was that a lot of people thought they should tell her how to eat. One neurologist told her to avoid all saturated fat, a recommendation based on the work of Roy Swank, a doctor at Oregon Health Sciences University who observed his MS patients faring better on low-fat diets in the 1950s. Now a dietitian, Anna raises an eyebrow at giving out decades-old nutrition advice. But at the time, she says, "it felt like a very general recommendation, just part of the protocol for new MS patients."

Saturated fat wasn't the only food Anna was told to avoid. Other doctors encouraged her to cut out all so-called inflammatory foods—the standard elimination-diet targets of gluten, dairy, eggs, corn, and soy. And her mother began to buy foods that were supposed to be good for Anna's brain: tofu hot dogs and salmon, for instance. One day she came home with a juicer and told Anna, "If you're feeling symptomatic, don't eat food, just drink vegetable juice!" She also got Anna a membership at the nicest fitness center in town.

"My mom had the very best of intentions," Anna says. "Especially in the first years after my diagnosis, my parents' entire focus was on me and my health. So she wanted to take really extra-special care of me. But it changed the way I looked at food for the first time." Anna had always been a rule-follower, but now she was also a teenager grappling with this thing that was going to change her life in ways she couldn't understand. Sparring with her mother about what she was or wasn't eating while trying to absorb the new recommendations from doctors and other specialists became a running theme of Anna's adolescence. And she didn't like that food had suddenly become this scary thing, full of land mines. "It felt like everything had the potential to hurt me," she recalls. And yet, like most teenagers, Anna wanted to think of herself as immortal, as somehow impervious to the worst implications of her diagnosis—even as she found herself frequently back in the hospital for more tests and treatments, or headed to school with an IV line infusing medication through her veins. "I just kept bucking against it," she says. "And sometimes now, I get a little tripped up, because I'll think, 'Gosh, I really wish I would have listened to my mom.' But also, I can't promise you that would have made any real difference. At the end of the day, I still have MS and I'm always going to have MS."

Anna says the constant stream of well-meaning but not always science-backed dietary advice continued throughout her college years. She worked with acupuncturists, naturopaths, and Ayurvedic practitioners. At one point, the acupuncturist Anna saw in high school introduced her to a monk who had recently flown in from China. "He didn't speak any English, and I spent several days with him, learning qigong and doing energy work," she recalls. "I was sixteen. It was weird." This history speaks to her parents' desperation—to the weight of that terrifying diagnosis and the need to try anything that might fight it. "We didn't want MS to be the final answer," Anna says. But she also recognizes the role of privilege in her family's ability to chase down so many unusual strategies. "If we didn't have the financial wherewithal to do any of these things, I would have had to just take my MS diagnosis and get on with my life," she notes. "In some ways, that might have been better for me."

By her twenties, Anna began trying to tune out the really fringe ideas, in part because she was laser focused on her career path. She graduated college in less than three years and had her master's done soon thereafter. "I kept thinking that I want to do this work while my body will still allow me to do it," she explains. Anna was drawn to nutritional studies in part because of her own experiences in the medical world. She decided to focus on eating disorders because her sister struggled with bulimia for several years, a condition that developed around the same time as Anna's MS and that was, in some ways, even more confusing for the family to navigate. "When my sister was at her sickest, I decided I wanted to become an eating-disorder dietitian so I could tell her that people get better," Anna says.

But studying nutrition also pulled Anna toward the same kind of thinking that had driven her mother to stock up on tofu dogs and kale—only now she was the one scrutinizing her own diet for failures. When they're learning how to use nutrient-analyzing software, dietetics students are often asked to provide "twenty-four-hour recalls," lists of everything they've eaten in the past day. Officially, nobody was grading Anna's lists, but "I was very aware that if you're a nutrition major, you should be eating in certain ways," Anna says. "So I falsified my twenty-four-hour recall. And that led me to a phase of tracking everything I ate pretty consistently. It wasn't even necessarily because I thought I would do anything with it. But something about school gave me permission to do this. And then it started to seem like something I was supposed to be doing."

Anna's tracking habits never progressed to the degree of

orthorexia that Christy or Emily struggled with. She began to learn about intuitive eating fairly early on in her career, and quickly embraced that approach, both for clients and for herself. She left one eating-disorder clinic because she was bothered by the amount of weight stigma she encountered there. "Patients with larger bodies were approached very differently than patients with smaller bodies. Their eating disorders weren't taken as seriously and they weren't addressed quite as humanely," she says. "It just didn't feel right to be treating patients so differently based on their size." Her next job was as an in-house dietitian for Monte Nido Laurel Hill, a residential eating-disorder treatment facility in Boston where the strategies adhere more closely to a "health at every size" philosophy. She is now the director of nutrition services for all Monte Nido facilities, in addition to running her own private practice.

But then, as Anna puts it, "my body came a-calling again, in a different way." Four years ago, her relapsing, remitting MS transitioned to secondary progressive MS. Anna no longer has periods of remission but instead lives with certain symptoms of the condition all the time, and can expect them to get worse. A limp so mild that she had barely noticed it for years has become more pronounced. On most days, she now uses a walker to get around her house, and a cane when she's out in public.

As Anna's health got worse, her determination to address the disease head-on increased. "I've really chased after not wanting this diagnosis to be the end of my story," she says. She saw a doctor who thought maybe all her troubles were due to undiagnosed Lyme disease, but several rounds of antibiotics failed to make much difference. Another decided she really had a yeast infection and prescribed a "*Candida* cleanse," in an attempt to reset Anna's system. In 2016, she read a book called *The Wahls Protocol*, whose author,

Terry Wahls, a clinical professor of medicine at the University of Iowa, has secondary progressive MS. On her website, Dr. Wahls touts her decades of academic medical training and notes that she has published "over 60 peer-reviewed scientific abstracts, posters and papers." But the original subtitle of her book was *How I Beat Progressive MS Using Paleo Principles and Functional Medicine.* "Right there, I mean, that's not a thing," says Anna. "You can't heal from progressive MS." But she decided to go on the diet anyway. (In later printings, Wahls edited her subtitle to tone down its promise. The book is now called *The Wahls Protocol: A Radical New Way to Treat All Chronic Immune Conditions Using Paleo Principles.*)

For four months, Anna ate nine cups of vegetables a day and sixteen ounces of liver and other organ meat per week. She didn't touch grains, dairy, or soy. The result was just like what had happened in high school, only even more severe. She couldn't go out to dinner with friends; every meal had to be planned out in advance. And organ meat made her gag. "I didn't look forward to eating at all. It was just this mechanical thing I had to do three times a day." All of which might have been worth it if she had started feeling better. But nothing changed. "I was heartbroken," she says. "This was supposed to heal me from this thing." And then she fell down the same rabbit hole as so many dieters before her: She blamed herself, not the diet. "It was 'Oh, my God, I can't do this right,'" she recalls. "I mean, Terry did this and she 'healed'! So why do I still feel like shit?"

In signing on to the Wahls plan, Anna says, "I really disconnected from my own truth and my own body." And that was frightening because Anna spends all day, every day, trying to help eating-disorder patients reconnect with their bodies and with food. One of her favorite parts of her job is sharing meals with her clients, helping them to rediscover the pleasure of eating. On the Wahls

diet, she couldn't do that. In fact, she couldn't even let her clients
know that she was on such a diet. The lists of banned foods and
the inflexible rules about when to eat posed dangerous ground for
someone with a history of such restrictive behaviors. And it
didn't matter that Anna was following the Wahls plan for health
reasons and didn't care about losing weight. "I was sitting there
with clients, talking about the importance of intuitive eating," she
says. "And here I am, doing something else."

Yet to Anna's surprise, most of her colleagues in the eating-
disorder treatment world didn't see her diet experiment as a professional conflict. She presented her situation to a panel of colleagues
whose job it is to help spot and navigate such issues. They were
unanimous in their support. "My peers kept saying, 'Take care of
yourself! Do what you need to do for you!'" she says. The panel
saw no problem with adopting extreme eating habits as a way of
managing a major medical condition, and on the surface, maybe
there isn't one. We all accept that a diabetic has to avoid sugar or
risk a dangerous insulin spike. A patient recovering from open-
heart surgery is unlikely to be prescribed a diet of cheeseburgers
and milkshakes. Violet needed her feeding tube because she would
have starved without it. There are times when the science around
nutrition and health really is black-and-white, when perhaps our
emotional connections to food should take a backseat to our
physical needs. But Anna knew deep down that what she was doing
with the Wahls diet wasn't that clear-cut. The dietary advice was
too extreme, the book's scientific premise too tenuous. It relies
largely on data from animal studies, much as Merrell's claims about
detox science are primarily extrapolated from in vitro research. And
many in the MS research community have spoken out against the
Wahls approach. What Anna was really doing on the diet was
fighting the reality of MS in her body. She couldn't "own her truth"
of how the diet was affecting her, because that would have meant

accepting that she was no longer an able-bodied person. And might never be one again.

A week after she stopped the diet, Anna flew to Las Vegas for a conference on eating disorders. The event was housed inside a large hotel and casino and there would be a lot of ground to cover each day. So Anna decided to rent an electric scooter for the first time ever. At first, it was fun to zip around at fifteen miles per hour. But on the second day, when the exhibit hall filled up with marketers and professional colleagues, Anna felt the air around her change. "Suddenly, people in my field weren't looking at me," she says. "Marketers weren't trying to sell me anything. People were talking about me instead of to me. I get that they didn't want to stare. But when someone consciously doesn't look at you, you become invisible."

The palpable discomfort around her disability was sharpened by the fact that eating-disorder professionals are, almost by definition, supposed to preach acceptance of all body types. And yet, Anna says, she was far more aware of people politely turning away from her at the conference than she was when she went out on the main casino floor, filled with regular Las Vegas tourists. "I don't know if it's because they've all been in a casino since nine a.m., or if they were somehow more accustomed to seeing different bodies," she explains. "But strangers there smiled and engaged with me like it was no big deal." One day Anna attended a presentation on body image during which the lecturer asked the audience whether anyone could share an experience of people appraising their bodies or noticing them differently. One woman talked about how she felt people's awareness of her body changing as she got older. Someone else brought up sexual orientation and transgender issues. "And here I am, in a fucking scooter, and we did not talk about ability," Anna says. "I was so upset. At everyone else, and also at myself, because I was too much of a wuss to raise my hand."

Anna decided it was time to "come out" as a person with a disability. She went on Christy's podcast and shared the story of her disease and her experience at the conference. "I no longer have the ability to not identify as a disabled woman and I no longer want to do so," she told Christy. "I'm not sure what I want to do with that. But I do not want anyone, anywhere, to feel the way I felt over those four days at the conference."

I include this part of Anna's story because it speaks to how widespread body anxieties are within the wellness industry—even among eating-disorder therapists, whom you might expect to be the most enlightened on these questions. And if therapists, dietitians, coaches, and other so-called diet experts are uncomfortable with body diversity, or perhaps feeling bad about their own bodies, we should consider how this skews the advice they give us about what and how to eat.

Social media is now awash in self-styled fitness and diet "experts"—usually bloggers, personal trainers, or enthusiastic home chefs who have no training or credentials beyond their passion for juicing and perhaps, their own personal experience of losing a significant amount of weight or otherwise transforming their bodies. And while educated consumers like to think that we're too savvy to buy into their messaging, there's a reason that Gwyneth Paltrow's GOOP brand has grown from "homespun weekly newsletter" to multimedia platform, with over 791,000 Instagram followers and an organic skincare line. Meanwhile her friend and fellow "influencer" Amanda Chantal Bacon didn't enter the game with Paltrow's celebrity credentials but has nevertheless racked up 202,000 Instagram followers for her multimillion-dollar Moon Juice empire, which sells $38 jars of edible "superherbs" that claim to boost stamina and control your cravings. Many of us find these well-packaged promises utterly seductive. Still, if the diet fails, we might think, "Well, next time I really should consult a trained professional like a doctor or a dietitian." But what if those professionals are just as susceptible to that diet-culture messaging?

"The health and wellness professions are very attractive to people who have their own issues with food," says Christy Harrison, the former food writer who struggled with orthorexia. "Or these issues can get stirred up by the profession, by the work." This can be a good thing. Now, as a dietitian, Christy says her own experience with disordered eating helps her feel empathy with her clients as they wrestle with how chronic dieting habits have started to take over their lives. But the draw of a professional life focused on food issues can also be dangerous. "I was a practicing registered dietitian at the time [when I had my eating disorder]," says Emily Fonnesbeck. "I've had to really let that sink in." Emily has issued public apologies on her blog for the bad advice she believes she gave clients during that time. But she's even more concerned about how often she sees a similar dynamic playing out within the industry. "I don't think my experience is really unique in terms of health professionals."

In eating disorder recovery centers, patients are encouraged to throw their scales away and stop obsessing over hitting unrealistic goal weights. But the clinicians treating them continue to use weight as a primary tool for assessing patient health; most use the body mass index scale, or the same "ideal body weight" metric that Christy learned from her dietetics textbook. Anorexics at inpatient facilities are weighed daily so their tube feeds or meal plans can be adjusted accordingly if they lose or fail to gain weight as expected; insurance companies also use these numbers to determine whether to keep paying for treatment. Usually, patients are shielded from seeing the results of those daily weigh-ins, but sometimes clinicians share the numbers, thinking that being able to tolerate a "healthier" weight is a critical part of recovery.

Meanwhile, patients with bulimia or binge eating disorder are also weighed as a way of gauging their progress, though for these folks, paradoxically, clinicians may be looking for weight loss as a sign that destructive overeating has abated. In her work in eating-disorder centers around New York City, Christy says, it's common to hear therapists bemoaning a binge eater's failure to lose. "They'll say, 'Well, she must still be using behaviors because she's not getting down to her ideal weight!'" she says. "Which makes no sense. That's the toxic diet mentality we're trying to get them away from in the first place."

Dietitians may focus on weight in part because they don't get any training in how to identify disordered eating patterns. Anorexia and bulimia might be covered in one unit of one class dealing with a bunch of rare food-related disorders. "It's all about calories, how to help people lose weight, how to calculate tube feedings, how to handle renal failure," says Christy. "Everything we learn is optimized for acute-care settings. There's a lot less emphasis on what everyday people are struggling with out in the real world." There's also no effort to screen dietetics students for eating problems, in the way that, say, psychology students are encouraged to be in therapy themselves. Yet several studies suggest that nutrition students have a higher prevalence of eating disorders than college students with other majors. One such study from South Africa found the risk to be double, with 33 percent of nutrition students displaying signs of eating disorders, compared with 16 percent of the rest of the student body. Even Steven Bratman, the doctor who invented the term "orthorexia," acknowledges in his book *Health Food Junkies* that he did so after diagnosing himself with the condition. "We all have our own stories about body image," Anna says. "And many of us are making recommendations to our clients that we don't follow ourselves." (I contacted the Academy of Nutrition and Dietetics to ask for their take on how eating disorders are included in dietetics curricula, as well as the research suggesting higher rates of disordered eating patterns among dietetics students. The organization declined my request for an interview.)

The bigger issue may be that even among clinicians who work on the mental ramifications of food, the belief is widespread that what a diet does to your body matters more than what it does to your psyche. That's why Anna's colleagues supported her decision to undertake a strict elimination diet. Emily has encountered the same kind of response when she speaks publicly about how her experience with the LEAP program and elimination diets fueled her orthorexia: disciples of such programs email her, aghast that she would be so dismissive of the real pain and suffering they see their clients experiencing when they eat the "wrong" foods. "I'm not saying that elimination diets never work for anyone or that they will always trigger eating disorders. But I do think we need to be very careful about putting vulnerable people on these plans," Emily says. "And you wouldn't think that would be controversial to say, but it is."

Consider the title of a piece written by Columbia University psychiatrists and published in the May 2016 *International Journal of Eating Disorders*: "Long-Term Weight Loss Maintenance in Obesity: Possible Insights from Anorexia Nervosa?" It's an odd analysis of the "remarkable parallels between behavioral patterns" of people with anorexia and the small percentage of dieters who do successfully manage to lose weight and keep it off, as tracked by the National Weight Control Registry. Both groups restrict their food intake for years; both have lower resting metabolisms and higher levels of ghrelin, a hormone that signals hunger, than control groups, and yet they are somehow able to override that biology to maintain a significant weight loss. But rather than asking whether this means folks on the National Weight Control Registry are perhaps also undiagnosed anorexics, the paper's authors see these similarities as fertile ground for future study—which, they say, could lead to "new or improved interventions" for people trying to lose weight. The journal published the piece under the heading "An Idea Worth Researching."

So even wellness professionals identifying the problems with obsessive "healthy" eating feel the need to toe the line, to obey a kind of nutritional hierarchy in which obesity is framed as the primary health threat, and all other eating issues are perceived as minor by comparison. And that leads to a certain hypocrisy: trying to treat a disease that's rooted in an obsession with body size by . . . tracking body size. Maybe when anorexics fixate on BMI, and orthorexics obsess over gluten, they aren't being crazy—they're just more honest than the rest of us. They know—because they're hearing it from everyone, even their mental health professionals— just how much these things are supposed to matter.

"I don't know what normal is, but I wish I was normal," says Karen Levy. She's a forty-seven-year-old TV executive and mother of three kids in Massachusetts, who estimates she's been struggling with what she calls "eating issues" since her parents took her to a diet workshop when she was nine years old. "I don't want to think about what I eat all the time. I want to eat what I want to eat," Karen says. But she can't ever seem to quite get there; she's too convinced that at "120-something pounds" and five feet, three inches tall, she needs to be thinner. "I don't look like I have an eating disorder," she says. "I'm more the type who can look at an apple and gain ten pounds." And so, she doesn't eat the apple. Karen doesn't eat many foods; if she bakes cookies for her kids, she might have one lick of batter, but won't eat a whole cookie. She cooks dinner every night so the whole family will sit down together. "But then I can't bring myself to eat the rice. It's so dumb."

But Karen's diet fixation isn't just about weight. She's also struggled with gastrointestinal issues since her teenage years. In college, she was diagnosed with irritable bowel syndrome and lactose intolerance. Medication didn't really help. "When your stomach kills, it contributes even more to the whole food thing," she tells me. "So I've seen a lot of doctors. Like, maybe every GI on the Eastern Seaboard." She's also seen functional-medicine doctors, acupuncturists, and dozens of other alternative healthcare professionals. Every time, she says, they run a battery of tests, "which tell me nothing." And then they put Karen on a diet. She's done gluten-free, dairy-free, and FODMAP, a diet developed for irritable bowel syndrome patients by researchers at Monash University, which also sells its own certification program, diet booklet, and testing protocol. The acronym stands for "fermentable/oligo-saccharides/ disaccharides/monosaccharide/polyols," all technical terms for various compounds found in an utterly bewildering list of foods: Cashews are forbidden, but almonds are fine; apples, pears, and mangoes are out, but grapes and melons are in. As a lifelong restricter, Karen has no problem following a set of food rules, though she grumbles about how much work is involved in parsing such complex lists. "You have to read every freaking label. But I'm religious about it, because if I'm going to do it, I'm going to do it right," she explains. She also admits, freely, that the diets never really work. "I learned on FODMAP that broccoli makes me gassy," she notes dryly. "It wasn't exactly an eye-opening revelation."

When I talk to Karen in May 2017, she has just come from seeing yet another new "quack doctor," this time a kinesthesiologist who has decided she has a hypothyroid problem and needs to work on her carb and sugar intake. "Now I have another person telling me to take something out, even though I know for my eating disorder, I should be putting things back in," she says. "I don't want to eliminate those foods from my life. But in our society, there are good foods and bad foods. That's just how it is."

I ask Karen whether she thinks her chronic stomachaches are related to her years of restrictive eating. "Oh, for sure," she says. "I only ate five hundred to eight hundred calories per day for a long time. That trashes your metabolism. And stress makes everything with your gut worse." But when she meets with a new doctor or wellness professional, she doesn't tell them the whole story. "I'll say I have eating issues and I work with a nutritionist," she says. Karen is a longtime client of Marci Evans, a Boston dietitian who specializes in disordered eating and also works with weight-loss-surgery patients. But she doesn't spell her history out further and she says they rarely ask. "I don't think they see a connection between 'eating issues' and putting me on their diet."

To the wellness industry built around the good food/bad food dichotomy, Karen is a list of physical symptoms—constipation, bloating, stomachaches—to be resolved once they finally find the true culprit in her diet, the bad-food bogeyman that's capable of wreaking such havoc. But the co-occurrence of eating disorders and digestive disorders is well over 90 percent, according to a 2010 analysis published in the journal *Neurogastroenterology & Motility*. A more recent survey of 185 Australian eating-disorder patients found that 23 percent had acid reflux, 27 percent were constipated, and 42 percent had irritable bowel syndrome. Restricting and purging can take a fairly obvious toll on digestion, and both patient populations have high rates of depression and anxiety, notes Evans, who sees many who are struggling with both issues. And the authors of the 2010 analysis noted that data on long-term eating-disorder patients shows that even once they begin to recover and gain weight, their GI symptoms can persist. They may have dieted themselves into another health problem. Which most doctors will want to treat by putting them on a new diet.

Karen, and so many like her, will go on that diet—even though she knows it's counter to her eating-disorder recovery goals, and even though she knows how unlikely it is to make her feel better. She'll follow the new plan and cut out a new round of foods, because she's still looking for what we all seek: A way to feed ourselves that makes sense. That feels simple and right. That doesn't make us feel guilty about everything we put into our bodies. We no longer trust ourselves to know this intuitively, and maybe some of us never did. So instead, we're searching for something external: an expert we can trust, a set of rules to follow, a literal recipe for how to develop this basic life skill. Many people within the wellness industry are searching for the same thing. But in the meantime, they're happy to sell us their new plan.

Three

Comfort, Food

For Eva Ingvarson Cerise, the pressure to raise a perfect eater began before she even conceived her. Eva was forty-two and had already had one miscarriage when she and her husband, Kirk, began trying to conceive again. Eva is a planner and a perfectionist. At the time, she created websites for a large financial firm in Los Angeles. She's also into the West Coast's alternative health scene, with its promises of a more holistic, natural lifestyle. She's not crunchy, exactly; she's part of a new trend of consumers who see their diet and lifestyle choices as a constant opportunity for self-improvement and empowerment. Healthy living offers a way for Eva to be her best self. And also, she wanted a baby. Badly.

So Eva did what she calls "the typical over-forty thing" of going to a naturopath and an acupuncturist to try to boost her fertility. By "typical," of course, she means for mostly white women of a certain social class in Los Angeles, which is more or less ground zero for the wellness industry and its obsessive relationship with food. Going on a special "fertility diet" doesn't seem strange there, because people are going on and off detoxes, cleanses, and other kinds of diets all the time. Eva wasn't surprised when both of her practitioners recommended overlapping versions of the same fertility diet, which they claimed would regulate her hormones by cleaning her liver with foods like beets, lemons, and artichokes. Eva's reproductive endocrinologist was "less strict," but nevertheless advised her to quit caffeine and alcohol and eat as healthfully as possible. Eva followed all their diet advice to the letter. "That was the Halloween where I ate prunes," she says. "That was fine."

But eating so meticulously had ripple effects. Even in health-conscious LA, Eva says she saw relationships with friends change because they could no longer share meals. She stopped accepting dinner invitations because restaurant menus were too daunting to navigate. And getting pregnant suddenly seemed less a shared project with Kirk and more a solitary mission. "I had to think about food all day long, every day," she says. "He was still eating like we always had. I felt very alone."

It took another miscarriage and three rounds of IVF, but Eva finally sustained a pregnancy when she was forty-four. "I know my age was the biggest factor in my fertility troubles," she says, "but I do think my diet could have contributed to the quality of my one good fertilized egg, and to my body's ability to sustain that pregnancy." And during the nine months that Eva spent growing her daughter, Annika, her obsession with eating perfectly only increased, especially in the first trimester, when all she could stomach was sour candy and crackers. "After my miscarriages, I felt like I had to do everything I could in order to keep this pregnancy going, and here I couldn't eat any healthy food," she says. "It was terrifying." But her morning sickness subsided in the second trimester, which she describes as "blissful." She walked, swam, went to prenatal yoga, and dutifully followed all of her OB-GYN's rules about pregnancy nutrition, which meant avoiding soft cheese, frozen yogurt, and salad bars, and eating no more than twelve ounces of fish per week to limit her mercury exposure. The diet was a little complicated, but also satisfying because everything was going so well. Then, at the end of her second trimester, Eva was diagnosed with gestational diabetes as well as a rare bone condition called transient osteoporosis of the hip. The doctors who diagnosed her osteoporosis wanted Eva to eat lots of calcium and protein. But her endocrinologist wanted her to limit dairy because it contains naturally occurring sugars that can interfere with blood glucose levels. Eva found she couldn't even put milk in her morning coffee without a blood sugar spike. "The two conditions completely worked against each other," she says.

Eva had gained eighteen pounds by that point in her pregnancy; in the weeks following the diagnosis of gestational diabetes, she lost seven. "They said nobody ever followed the guidelines so fully," she says. "But I'm a rule follower, especially when you tell me high blood sugar might harm my baby!" Her doctors were pleased, but Eva was unnerved by the sudden weight loss. And in the last few weeks of her pregnancy, the baby stopped gaining weight as well. She began eating every two hours, alternating bland protein shakes with three eggs at a time or several handfuls of almonds, and her maternal-fetal medicine specialist told her to ignore any restrictions on dairy and carbohydrates and eat lots of high-fat foods. But at the endocrinologist's office, a nurse told Eva that eating that way would risk her baby's health. "It felt like both specialists thought following the other's advice would harm my baby, and I was caught in the middle," says Eva. "Nobody talked to each other or helped me come up with a 'safe' diet, so I was left to figure it out on my own." She decided she could eat yogurt and bread in the afternoons, when her blood sugar was more stable, as long as she tested it frequently, and that seemed to work. But still: "I was so frustrated," she says. "It felt like I was jeopardizing my pregnancy at every meal."

Eva's food anxieties are in large part, the result of her privilege; not every pregnant woman can splurge on acupuncture and fertility teas. But every socioeconomic group today is subjected to an onslaught of messaging around the importance of prenatal nutrition. We're told that whether we eat enough protein or too much sugar will directly impact our baby's development in utero, but also whether she grows up to like vegetables or be fat or get heart disease. And women aren't encountering this pressure in a void—it piles on top of whatever food anxiety they were already incubating. Yet few prenatal-care providers talk about what happens when you start off frustrated by your weight for primarily aesthetic reasons—an already fraught situation, to be sure—and then add on the overwhelming responsibility of building a healthy baby with every bite you take. For women who are unhappy with their weight before they conceive—and that's two thirds of us, most of the time, according to a 2016 study published in the journal *Body Image*—it's difficult to transition to dispassionately assessing their weight gain during pregnancy as a neutral marker of health. Instead, it becomes evidence of whether we're getting pregnancy "right." And that's only the beginning. Our anxieties about weight and diet continue to filter into how we think about feeding our babies from birth on into early childhood, when everything diet culture wants us to do seems to contradict how the little people we're feeding really need to eat.

"Food was not fun when I was trying to conceive or when I was pregnant," Eva says now. "But I had this crystallizing purpose. I just had to have this healthy baby. I would have done anything."

And all the diet tweaks and restrictions still felt temporary. She knew her gestational diabetes would disappear when Annika was born, and the doctors expected her osteoporosis to resolve as well within three to fifteen months after she gave birth. Annika was born on December 4, 2015, weighing seven pounds, seven ounces. But her cry was weak, she spat up every time she tried to nurse, and in her first forty-eight hours, she failed to pass any stool, which can be a sign of intestinal problems. Eva's hospital transferred the baby to the neonatal intensive care unit, where she spent the next week undergoing various tests and treatments. "It was only a week in the NICU. I know how lucky I am," Eva says. "But it was also awful, because I just knew, in my gut, it had to be related to how I had nourished her during the pregnancy."

Eva's one source of relief was the thought that at least eating could go back to normal. She could breast-feed and nourish Annika properly by eating a wide range of foods, and she could finally indulge in everything she'd been denied for nine months. And at first, it seemed as if she was right. Since she had lost so much weight during pregnancy, she immediately began enjoying bread, chocolate, ice cream, and everything else that had been on the restricted list. "I felt like I was owed," she says. "I had earned the right to eat whatever I wanted." Annika finally pooped and the family went home. And then, everything got worse.

When Annika was six weeks old, she developed a rash and started screaming and pulling off the breast every time she nursed. Her stool also changed, becoming "mucusy" in consistency. This was especially alarming to Eva after Annika's rocky first week. She worried that the two digestive problems were related and that her baby was somehow still suffering ill effects from Eva's weird prenatal diet. Eva and Kirk called the pediatrician three times, and were told each time that Annika was probably "just fussy." Finally, after two weeks of this, they were told to bring the baby in. The doctor tested Annika's stool and found blood. She explained that when Eva ate certain foods, traces of those foods were absorbed into her breast milk, which then would inflame Annika's intestines and make her bleed. "Basically, she was reacting to the proteins in most of the foods I was eating," Eva says. Even if her prenatal diet had no lasting impact, every bite Eva took now seemed like playing Russian roulette with her baby. And there was no way to know for sure which foods were causing Annika's inflammation.

On her pediatrician's orders, Eva tried to narrow the list of suspects by cutting out the eight most common food allergens: dairy, soy, tree nuts, peanuts, seafood, corn, eggs, and fish. Still, the bloody stool continued, so Eva added wheat to the list. Annika seemed to react to almost anything Eva ate, even kale. "I was terrified to eat, terrified to nurse her, terrified to change a diaper and find more blood," she says. Her pediatrician seemed stumped, so Eva began her own research, consulting a naturopath and joining Facebook groups where mothers of babies with food intolerances compared notes and strategies. Elimination diets are in constant rotation on these groups; Eva dismissed a popular plan that would have required her to drink straight chicken broth for several days. "That felt way too intense." But she decided to try one that would cut her down to just seven foods: chicken, quinoa, turkey, zucchini, sweet potatoes, apples, and pears. "I was also allowed to have olive oil, salt, and pepper. I did that one because my naturopath said these were all low-allergy foods that would keep me satiated," she says. "I mean, these are all crazy, extreme diets, but for women whose babies have this, it's the only thing that gets them to baseline."

Support groups like the ones Eva joined are closed and often require mothers to fill out extensive waivers in order to participate, both because sensitive medical information is discussed and because these mothers are used to encountering a fair amount of judgment and shame. Eva understood because she was experiencing the same thing. "My immediate family was supportive, but most other people told me I was being crazy and extreme, and kept suggesting that I just switch to formula," Eva says.

I also ask Eva why she didn't consider formula. It's clear both that she is sick of hearing the question and also that she never really considered formula a viable option. Annika had vomited when she was given some during her NICU stay. "I didn't want to try it again, especially since most formulas are dairy- or soy-based, which we now knew Annika couldn't have. So we'd have to import this European allergen-free stuff," she explains. "Also, talking to those crunchy moms online, they all said breast milk is better than formula even if she's reacting, because she's getting that better variety of nutrients from me."

There is an obvious flaw in this logic: If Eva could have found a formula that didn't trigger Annika's internal bleeding, surely one could argue that that would have been more healthful than depending on the breast milk that did? And yet—her thought process makes me remember Violet's heart surgeon saying, "I prefer breast milk," and my corresponding despair as my supply dwindled. If "breast is best," as new mothers are told relentlessly, then doesn't a sick baby need the very best of all? Even when breast milk is the apparent problem—as it was for Annika, as it was for Violet when the act of breast-feeding became too physically taxing and traumatizing—it is hard to shake the aura of "liquid gold," or to feel as good about opening up a can of Enfamil. "I was committed to breast-feeding because it is important for a baby's gut health and I had tremendous guilt over my daughter's gut issues," says Eva. "It's complicated emotional stuff, but she will breast-feed for as long as she wants."

But even as she pressed on, Eva wondered about the nutritional value of her particular brand of breast-feeding. "I worried all the time because here I was not nourishing myself," she says. "I know the baby sucks out the best I have, but what was I really giving her?" It's a valid question. In 2004, the National Academy of Sciences appointed a committee of ten pediatricians, dietitians, and research scientists to study the health benefits of breast milk and formula. They published a 220-page evidence review in which they concluded that "breastfeeding is the standard by which all other infant-feeding methods should be judged." But they also pointed out: "Human-milk composition varies considerably among and within individuals over time." This variation can include energy density, which may range from 15 to 24 calories per ounce, as well as the concentration of some essential nutrients. Up to 30 percent of breast-fed infants develop an iron deficiency (double the rate of formula-fed babies) and they may also be more likely to develop vitamin D deficiencies because human milk contains low levels of vitamin D, while formula is fortified. (Accordingly, the American Academy of Pediatrics recommends supplementing all breast-fed infants with vitamin D daily.)

"What a mom eats will absolutely impact her breast milk, and if you don't have enough calories, your body will shut down production," says Christie del Castillo-Hegyi, an emergency physician and co-founder of the Fed Is Best Foundation, a nonprofit advocacy organization that studies the relationship between breast-milk production and newborn brain injury, jaundice, dehydration, and hypoglycemia. "That's why we don't produce milk on the first day after childbirth. Labor has depleted you and you need to recover." Castillo-Hegyi says there is no good research on how elimination diets might affect a mother's milk supply, but she is concerned that "Breast is best" campaigns often under-inform women about the potential risks of exclusive breast-feeding. "Too many pediatricians' offices have this culture of reassurance, where everything is fine until something really bad happens," she says. "But any neonatologist will tell you that they admit a handful of babies every week who become dangerously malnourished when their mothers exclusively breast-feed despite insufficient milk supply."

On Eva's elimination diet, breakfast was a sweet potato covered in a quarter cup of olive oil; lunch was quinoa or plain turkey flavored only with oil and salt. She was supposed to eat that way for only a week, but she ended up doing so for much longer because being back at work full-time meant that Eva had to pump most of Annika's milk. On any given weekday, Annika drank from the stash that Eva had been faithfully stockpiling in her freezer. It took a month to replace all Eva's frozen milk and get to a place where she felt the diet was beginning to work. Slowly, Annika became less fussy, although her poop stayed liquid in consistency until she was nine months old and steadily eating solid foods.

At her nine-month checkup (by which time Eva had reintroduced many, though not all, foods), Annika's weight had dipped from the 50th to the 30th percentile, but her pediatrician said that kind of shift is common once babies start to crawl. She has otherwise continued to thrive and meet milestones; at age two, the worst of her intolerances are behind them though Eva still avoids feeding her wheat and lentils. Annika herself enjoys eating a range of solid foods, which she took to without much fuss. But for Eva, the experience has transformed how she thinks and feels about food. "I no longer eat for pleasure," she says. "Eating has become this constant quest to get the 'right' nutrients to nourish my daughter."

Choosing to breast-feed often means choosing to continue considering your own diet through the prism of how it nourishes your baby. How do you space alcohol consumption with nursing sessions? Is the baby crying because she's cranky, or because she's reacting to something in your breast milk? Are you getting enough of the elusive "variety" in your diet to expose your baby to the myriad of tastes she needs to ensure she grows up to like healthy foods? Choosing formula, as I quickly learned, doesn't absolve you of these pressures; if anything, it requires an even steelier commitment, since the choice is likely to be questioned by everyone you meet. And anxiety around these decisions appears to be universal: 90 percent of American mothers said that breast-feeding was best, even though only about half of them did so for six to twelve months or more, according to a 2015 survey of 13,519 mothers in ten countries performed by researchers for Lansinoh, a manufacturer of breast-feeding accessories such as nipple cream and breast pumps. It's important to note the probable bias of a survey about breast-feeding performed by a company that promotes and profits from breast-feeding. But it also seems fair to extrapolate from those numbers that at least some of us who stop breast-feeding sooner than we have been taught to consider ideal struggle with that decision.

As a result, prenatal and postnatal nutrition has become a big business: La Leche League International, once a grassroots collective of seven Catholic mothers, now has branches in seventy-seven countries and raised $1.2 million in revenue in 2015. And Baby-Center.com, a website owned by Johnson & Johnson, reports a monthly audience of 45 million parents. Which means even the most laid-back new mom is subject to constant messaging about food from the media as well as from her pediatrician, friends, family, and the Facebook mommy groups she joins in an effort to find comrades.

Weight gain is considered an important indicator of prenatal health because women who don't gain enough tend to have smaller babies, and babies with low birth weight (defined as less than five pounds, eight ounces) are also more likely to spend time in the neonatal intensive care unit. They're also more prone to have problems with their hearts, brains, lungs, and intestines, though some researchers, including Allen Wilcox, M.D., Ph.D., an epidemiologist who studies fetal development for the National Institutes of Health, have noted that the relationship between weight and such conditions is likely more correlative than causal. "Birth weight is one of the most accessible and misunderstood variables in epidemiology," Wilcox wrote in a 2001 evidence review published in the *International Journal of Epidemiology.* "Methods of analysis that assume causality [between birth weight and health outcomes] are unreliable at best, and biased at worst."

And most public health messages to pregnant women today focus on warning women against gaining too *much* weight rather than too little: "Like it or not, eating for two isn't a license to eat twice as much as usual," warns a MayoClinic.org article just below the headline "Here's Why Pregnancy Weight Gain Matters." A similar piece on BabyCenter.com admonishes the reader that "you probably don't even need any extra calories in your first trimester." To be fair, delivering a baby of "high birth weight" (above eight pounds, eight ounces) is associated with an increased risk for birth injuries, as well as future risk for diabetes, autism, and other health problems, but here again, birth weight has not been proven to be a root cause of those conditions.

Eva says she was "comfortable with eating," but "not so much with my weight" before she began that first fertility diet. "I liked to cook a lot, and try new restaurants," she says. "I was fairly adventurous." And cooking together was a pre-baby bonding activity for Eva and Kirk. But "I had the typical American-girl guilt any time I ate french fries or dessert," she admits. And she then adds, offhandedly: "I did have an eating disorder when I was fourteen, but that was more about control than thinness. Just your typical Type A Connecticut girl stuff. I grew up in a wealthy town, so this was my rebellion." The eating disorder lasted into Eva's twenties, but she says it was episodic, not chronic. Once a year or so, she'd get stressed out about life and start skipping meals. She'd drop ten pounds and everyone would tell her how great she looked. But really, she was too scared to eat. "I'd get so anxious that any food made me vomit," she recalls. "So I ate very little during the school day because I was scared to vomit. And I also became a vegetarian. It was all tied into my good-girl rebellion." Finally, in her twenties, Eva found a therapist to help her work through her social anxiety, and began to deal with her eating issues as well.

Eva doesn't connect her past disordered eating patterns with anything she's been through in the past few years. "I do stick with eating rules strictly, whether those are the rules I set for myself when I had the eating disorder, or rules imposed by a doctor. I don't cheat," she says. "But it feels different now because I have different skills to cope with my anxiety. I don't look at food as bad anymore." But she acknowledges that, good or bad, food—and planning what to eat, when to eat, and how it's likely to impact her health and Annika's—still occupies a tremendous portion of her waking hours. Because it's not just about Eva anymore. By the time a baby is old enough to begin expressing opinions about what's for dinner, a mother has often spent the best part of two or three years assuming total responsibility for a baby's nutrition and therefore, his or her health, growth, and mental and physical milestones. How a baby or young child then takes to eating on his or her own is yet another litmus test of how good a job you've done. If your toddler embraces kale smoothies and edamame, you've won. If not, you have only yourself to blame.

It's almost five p.m. on a January Friday in Brooklyn's hip DUMBO neighborhood, and Kate, a cantorial soloist and full-time parent, is trying to figure out dinner. Kate (who asks me to use only first names for her family) has actually spent much of the day thinking about dinner because tonight is the Jewish Sabbath, which the family marks each week with a ceremonial meal. A duck that marinated all morning in soy sauce and garlic is now roasting in her stainless steel oven alongside broccoli, cauliflower, and butternut squash. "Cooking duck makes me feel very fancy but it was actually just cheaper than the chicken," she tells me. Vintage silver candlesticks, plus one covered in glitter by her three-year-old daughter, Malka, stand ready on her marble dining table, a pretty little oasis amid crumbs from earlier meals, a few pieces of kid artwork, a sticky copy of *Jamberry*, and other detritus of family life. The sun is setting, so all we need now is for Kate's husband, Simon, to come home from his job running a midtown hedge fund so the family can begin Shabbat. But Simon has just texted that he's running late, which means Kate needs to fend off Malka and her one-year-old sister, Vivie, who are both starting to melt with hunger and whatever else it is that makes small children everywhere fray so desperately around the edges at five p.m.

"Mama, can I have milk?" Malka asks. She and Vivie are both big for their ages, with their father's solid build and their mother's infectious grin. Malka, who longs to be a ballerina, pirouettes around the living room, wearing only pink underpants. The girls are fresh out of the bath and she has not yet agreed to be clothed again.

"First let's have some cheese and crackers," says Kate, sliding over a plate of organic Ritz knock-offs and sliced cheddar. Malka takes one bite, then jumps down to show off her toe point. "Look, Mama! Feet!" she demands. "Feet!"

"Yes, feet! Amazing!" says Kate. She scoops Vivie into her high chair and serves her some leftover roast eggplant and lamb, which she's pulled out as a kind of appetizer for us all.

"I don't like eggplant," says Malka, observing Vivie. "Mama, I want my milk!" Kate brings her a previously discussed cream-cheese sandwich. Malka again eats one bite. "I don't like it."

I am confused as to why Kate isn't pouring a glass of milk, but then I realize: That's not the milk Malka wants. At age three, she's still nursing—something Kate is conflicted about. She's fine with it at bedtime, but tries to stave off the requests before dinner in what has become an all too familiar pattern. "We eat like, ten dinners most days," she tells me as she heads back to poke around in the fridge. In addition to the nursing debate, there is the meal-time chaos that most parents can relate to: Both girls are hungry earlier than she or Simon normally eat, and Malka, in particular, is a kid who never specifically says she's hungry, but turns into "a miserable little tyrant" when she doesn't eat enough. "I'm constantly trying to make sure I remember to feed her enough so we can avoid the tantrums," Kate explains. And that makes it hard to insist that Malka finish what she's started. "I just want to get something in her."

Kate offers granola and regular milk in a bowl, which Malka accepts. But then she stubs her pointed ballerina toe hard on the kitchen radiator. Tears ensue. The granola sits forgotten, soaking in its milk. A few minutes later, the cycle repeats. Malka rejects offers of a quesadilla, a hard-boiled egg, and an Amy's organic pizza pocket before choosing Cheerios, which leads to a negotiation over whether she can have Honey Nut Cheerios or the regular kind.

"These are too sweet, they're just for grown-ups," says Kate, tucking the coveted box away in a crowded pantry cupboard. Vivie placidly eats fistfuls of ground lamb and eggplant, watching as Malka melts to the floor in a classic three-year-old tantrum, feet kicking and fists pounding. "But I had it last week! I want it! It *is* for kids!"

They settle on mac and cheese, which Kate swiftly heats up and deposits on the table. Vivie grabs a fistful. Malka, again, takes one bite and zooms back to the kitchen, where Kate is now starting to carve the duck. "Mama, I want my milk! I didn't have my milk yet!"

"Why don't we go ahead and bless the challah and then you can eat that?" Kate says. "We can bless it again when Daddy gets home."

A few minutes later, Simon arrives. He's a sweet, smiling father and the girls are ecstatic to see him. They dissolve into giggles when he pops in and then immediately pops back out of sight, playing peekaboo around the corner of the loft. Kate puts dinner on the table, turns off the overhead lights, and lights the candles. Around us, the lights of the city sparkle through the apartment's enormous windows. Malka and Vivie gaze into the candles as Kate, who has a gorgeous voice and has performed in her childhood synagogue since the age of fifteen, sings the blessings. There is no whining. It is a moment of peace and wonder, the end to another chaotic week.

It is not, alas, the end of Malka's mealtime indecision. As Simon serves himself duck, he cleverly piques Malka's interest: "Oh, I don't think this duck is for you. . . . It's just for me!" She immediately demands her own duck, but then ignores it. "I want my milk, Mama! You said I could have it after I ate my mac and cheese!"

Kate has to acknowledge defeat. "You're right. I did say that. Okay, ten seconds." Malka hops off her own chair and onto Kate's lap, and nurses for a moment. "That's enough, Malka. You can have more at bedtime." But then Vivie begins to fuss; she also wants a turn. Kate takes a few hurried bites of her own dinner and lifts Vivie out of her seat. Simon clears away the mountain of food— mostly Malka's discards—that has piled up in front of the baby. When Malka sees Vivie nursing, she is displeased. Within seconds, both girls are crying. Kate calls it. "Okay. Bedtime!" Shabbat dinner is over. The duck that took hours to prepare is abandoned after fifteen minutes.

If you've never met her, Kate, age thirty, could be a little bit easy to dislike. She lives in a stunning DUMBO loft with expansive river views. The building was once a cardboard-box factory, but now features gleaming open-concept kitchens, high ceilings, and marble bathroom vanities. Kate's unit has three oversized bedrooms, but the whole family sleeps together in one king-sized floor bed, under a vintage chandelier. Malka and Vivie were both born in that bed, under the supervision of a doula and a midwife. And because of Simon's success, Kate doesn't have to work any more than she wants. Most of her days are spent with her girls, both of whom she has breast-fed for far longer than the American standard of less than twelve months.

Like Eva's obsession with acupuncturists and inflammation diets in California, this combination of privilege, chic aesthetics, and alternative lifestyle choices is often put together into a stereotype of white urban motherhood most frequently situated in Brooklyn or Los Angeles, though you'll also encounter it in cities like San Francisco and Seattle, and see it skewered on sitcoms. There's an expectation that an educated stay-at-home or work-from-home mother in one of these places will also buy only whole, unprocessed, organic food, enroll her children in enriching activities such as pottery for two-year-olds, have definitive views about sleep training, and perhaps be suspicious of vaccines. Mothers like Kate are supposed to do lots of yoga and maybe write a lifestyle blog with recipes for sugar-free birthday cakes. But Kate doesn't do either, and it's not clear where this stereotype originated. It may have begun with Gwyneth Paltrow and GOOP, her much-parodied lifestyle newsletter in which she promotes organic, macrobiotic cooking alongside $400 bedroom slippers. Another factor has been the rising popularity of attachment parenting, a philosophy espoused by La Leche League, the well-known pediatrician William Sears, and many celebrities. Attachment parenting encourages mothers to wear their babies, sleep with their babies, and breast-feed exclusively, on demand and for as long as possible. Modern food culture also plays a role: as we become more aware of ethical farming practices, toxic-chemical exposures, and our carbon footprint, upper-middle-class consumers have become more willing to pay a premium for organic, fair trade, and locally sourced food, especially for our kids. And the food industry has been delighted to market to these demands. My goal here isn't to shame parents who practice extended breast-feeding or want their kids to eat only organic; it's to understand why these beliefs about kids and food are so powerful—even as they are also often problematic.

Kate, in person, is far more thoughtful, irreverent, and complicated than the prosperous-white-urban-mother cliché. She doesn't present any of her choices as dogma, just as what's working for her in that particular moment as she wings her way through what she affectionately refers to as "the shitshow" of raising small children. And she resists being labeled. "Everyone I know gives their kids sugar, and most moms don't breast-feed for extended periods of time," she says. "I keep waiting to meet the hippie Brooklynites I read about in articles. Maybe they're in Park Slope?"

But Kate is also aware that her tendency to make unconventional life choices requires a degree of ambassadorship. "There's pressure to have had a perfect experience with it because you're representing this whole group," she says. "But I'm just not a very nostalgic or romantic person." When it comes to extended breast-feeding, Kate is glad to have had it work out so well for both of her daughters, but she's honest about the parts she doesn't love. With Malka, in particular, things started out rocky. Cracked, bleeding nipples, a painful latch, lots of crying from both mother and baby. In an email she sent me when Malka was a few weeks old, Kate described herself as a "crying, hormonal wreck," but she was nevertheless determined to keep at it. Her mother is a lactation consultant and a longtime member of La Leche League. "If I do formula, my mother will probably kill me. And then she will take Malka and bring her to a young, fertile wet nurse," Kate joked.

The pain got better, and Kate stopped needing to be in a certain position, with a certain pillow, to nurse successfully; she became a mother who could nurse lying down, standing up, on the subway, while eating a sandwich. And Kate assumed that weaning would happen naturally when they were both ready; instead, Malka's feelings about nursing seem to be intensifying as she gets older. "Malka has never, ever turned down an opportunity to nurse, and especially when she feels a little wary, like around new people, she'll ask to nurse even more than usual," Kate explains. "Meanwhile, Vivie generally nurses less during the day than Malka, and will often prefer something else, even when I offer. So now I know it's a personality thing and definitely more about comfort than calories."

And so, at age three, Malka is still breast-feeding, with no sign of weaning in sight. Yet the rest of her diet consists almost entirely of the high-sugar, refined-carbohydrate-heavy processed foods that the mothers in Kate's purported tribe are supposed to decry. The romanticization of extended breast-feeding tends to go hand in hand with a certain kind of nutritional ethos that makes very little room for mac and cheese and Honey Nut Cheerios. But the reality, in Malka's case, seems to be that the one has begotten the other.

"I think I was really very smug about Malka's diet," Kate says when she considers how they got here. In that first year, in addition to devoting herself to exclusive breast-feeding, Kate also read up on baby-led weaning, in which babies skip over purées and are instead fed only table foods, in sizes they can manage—another

concept pioneered by Kate's own mother and now embraced by
crunchy yoga mothers all over Brooklyn's elite neighborhoods. Kate
was careful to wait until Malka was sitting up and developing her
pincer grip before introducing foods, and then made sure to offer
a wide range of flavors and textures. "Other people would com-
plain that their toddlers wouldn't touch vegetables and I would just
feel so self-righteous," she recalls. "Like, I just gave my baby asparagus and she loved it!" Kate attributed her daughter's sophisticated palate to their breast-feeding success, because she had read
that breast-feeding exposes babies to more varied flavors than formula does, and she herself ate a pretty varied and healthy diet. "I
thought I was doing everything right."

There is an extensive body of research to back Kate up on this.
Julie Mennella, Ph.D., a biopsychologist at the Monell Chemical
Senses Center in Philadelphia, has been studying how babies learn
to eat based on the flavors in their mother's milk and amniotic fluid
since the 1980s. This phenomenon of "flavor learning" was first
documented in animals. Flavors that dairy cows eat, or even just
inhale, show up in their milk. Rabbits who are fed a diet rich in
juniper berries have offspring who seem programmed to choose
juniper berries when they're first learning to forage for their own
foods. Mennella conducted a series of blind sniff tests and taste tests
on human breast milk, which revealed that flavors like garlic, mint,
cheese, and alcohol are detectable if the mother has recently consumed them. In a later experiment, she demonstrated that the
human babies of mothers who eat carrots during pregnancy or
while breast-feeding are more likely to enjoy carrots as one of their
first foods. "Prenatal and postnatal flavor learning appears to be a
fundamental feature of all mammals," says Mennella. This makes
sense: Babies of all species learn from their mothers how to dis-
cern what's safe and nutritious to eat. And they don't have language
yet (or in the case of the cows and rabbits, ever). So we've evolved

to learn flavors in order to imprint our young with the information they need to feed themselves.

In my first conversation with Mennella, which took place around the time that Violet was beginning to eat her first full meals, I told her how surprised we were that Violet gravitated toward strong flavors like chicken tikka masala, spicy pasta sauces, and lemons, which she would happily chomp on whole. But then I remembered how Dan had complained that I put too much lemon on all our salads whenever I cooked during pregnancy; lemon was one of the few foods I craved. When I went nine days past my due date, I ate spicy curries and sauces almost daily in the hope they would jump-start labor. "See, that's it right there!" Mennella said excitedly. "Even without breast-feeding, she learned her early flavor preferences from you; she tasted all of that in your amniotic fluid."

The health implications of Mennella's work are far-reaching, because in our modern food culture, babies don't learn which berries are safe to eat in the wild. They learn the flavors of a modern diet from their mothers, and whether that diet is full of whole grains and vegetables or fast food and soda could, Mennella and others argue, impact the foods they gravitate toward as they grow. "We've got to understand the taste world of children in order to get them off to a good start," says Mennella.

But back to Kate, who, like most of us, doesn't spend a lot of time reading biopsychology papers and instead makes decisions about what to feed her kids that are informed by her own instincts and life experiences. In her case, those experiences started with a childhood in a rural part of New Jersey, where her mother, whom Kate lovingly describes as a "crazy hard-core health food nut," fed the family an entirely organic, locally sourced diet back in the 1980s before such things were trendy. Kate's dad has type 1 diabetes, so healthy eating was of paramount importance to the family. "This organic co-op would deliver all our food on a tractor-trailer once a month," Kate recalls. The family also had a huge garden. There was no junk food or fast food, not even restaurants of any kind; Kate remembers when her brother won a reading contest at the library and the prize was a Burger King gift certificate. "It was a huge deal because we had never been before," she says. "And I was probably twelve."

Kate doesn't view her mother's way as gospel. As a teenager, she began cooking for herself because she just didn't like her mother's food: "I liked bread and melted cheese," she says. "It wasn't meant as a rebellion. I just prefer fattier, richer foods with a lot of flavor and that's not really how my mom cooked. My favorite thing about being an adult is that I get to eat whatever I want now." But in a weird twist of fate, Simon is also a type 1 diabetic, which means Malka and Vivie have about a 1 in 17 chance of developing the condition as well, according to the American Diabetes Association. So Kate cooks family dinners that Simon can eat, like the duck and vegetables, and she initially steered clear of refined carbs when choosing foods for Malka. She felt proud when her new eater snacked on seaweed and broccoli.

Then, when Malka was around eighteen months old, she "just shut it down," as Kate puts it. She stopped trying new foods. And then she stopped liking foods she had previously liked, until all she ate was Kraft macaroni and cheese, Amy's organic pizza pockets, and eggs. Meals became a battleground, with Kate trying to persuade Malka that she still liked broccoli, and Malka spitting out every bite. Malka also wanted her food to look exactly the same every time. Eggs would be rejected if they carried a single fleck of visible black pepper. Chicken had to be scraped to remove all excess flavor. "She has developed every habit that I always heard about, but had attributed to poor parenting," Kate says. "You hear these stories and think, 'Oh my God, don't cater to all their demands; just say this is what's for dinner and let them go hungry if they

don't want it.' But I found I couldn't really follow through on that."
Malka's tantrums were formidable. Kate was pregnant again, and
Vivie was born when Malka was two. "It all went totally to hell,"
Kate says of trying to manage mealtimes with a newborn and a
toddler. A year later, Malka still eats pizza pockets or mac and
cheese for almost every meal. "I don't even offer the foods I make
for the rest of us anymore. I have just totally given up," Kate says.
"It's so embarrassing and so frustrating. I feel like I failed, and also
like, 'Come on, kid. What's going on?'"

Kate's annoyance stems partly from the fact that Malka will
eat just about anything that's offered at the preschool she attends
on weekdays from nine a.m. to two p.m. Her behavior is also much
better there. "It's made me realize that she's a kid who thrives on
routine and structure and I need to be giving her more of that at
home," Kate says. But she also feels a little let down by all the
people (including her own mother) who told her that she was doing
everything right in terms of breast-feeding and early flavor exposures. At a recent play date, a friend began waxing poetic about
her son's love of sushi. "I was just like, 'That's great, but I straight
up hate you,'" says Kate. And then the friend began to muse on
how her son's eclectic palate must be the result of the thoughtful
approach she took to feeding him, with unlimited breast-feeding
and frequent early exposures to a diverse yet healthful diet of table
foods. For Kate, it was like talking to her former self. "I mean,
come on. I did all of that! My kid should love sushi!" she tells me
while peeling hard-boiled eggs for the girls' after-school snack;
within minutes, Vivie will have crushed hers into the rug, while
Malka will begin lobbying for something different. "Meanwhile, I
get sad watching other kids eat a clementine because I realize she's
never even tried one. And that's a sweet, delicious fruit. That's supposed to be easy to like."

If early childhood nutrition can program your future health, then kids like Malka might be in trouble. But there is an alternative understanding of these children, which is that maybe they aren't contradicting Mennella's findings or developing early sugar addictions or otherwise going off the rails with food. Maybe they're progressing along a totally normal developmental trajectory—and have now hit a stage that is just as well documented in science as early flavor learning, but much less understood or even much discussed by parents in Kate's circle. Leann Birch, Ph.D., a psychologist who studies childhood obesity at the University of Georgia's College of Family and Consumer Sciences, was one of the first scholars to investigate how humans learn to eat in the first years of life. In 1982, she published the first in a series of papers describing how the initial period of flavor acceptance is followed by a stage of neophobia, a "fear of the new," which can last for much of early childhood. The paper cites research done on rats in the 1970s, which identified the presence of neophobia, as well as its gradual reduction; rats became more willing to eat a new food after repeated exposures. Birch herself later documented the same phenomenon in human children, noting that humans take much longer than rats to accept new foods; it took around ten feedings for infants and preschoolers in her studies to embrace a new vegetable—and that number may feel like a conservative estimate to many of us.

So why do almost all young eaters, even the previously adventurous ones, inevitably go through a neophobic stage? "This response may seem maladaptive, because omnivores need variety in the diet and the young child must learn to accept at least some of the new foods offered," Birch writes. "However, this need for variety must be weighed against the fact that putting something new in the gastrointestinal tract is a risky business." From that perspective, a toddler's disdain for clementines or kale can actually be understood as essential to the survival of a species that has only had the luxury of a safe food supply for the last two hundred–odd years (and even then, only in certain parts of the world). Prior to that, discerning eaters were far less likely to pick the wrong berries or eat too much spoiled meat. This understanding of neophobia is also a hopeful one, because, Birch emphasizes, this stage is temporary: "Neophobia does not reflect a fixed dislike for a new food, but a transitory one that may be altered via subsequent food experience," she writes. "The view that the neophobic response is normal and adaptive also implies that when children reject new foods, they are behaving normally, and should not be labeled as 'finicky' or 'fussy eaters.'"

This last part often gets lost as parents panic at the onset of neophobia and begin to reach for such labels. So if all kids must pass through this developmental stage, how many of them actually wind up as true picky eaters? Estimates vary wildly. Fifty percent of two-year-olds were identified as "picky" in a 2004 survey of 3,022 children published in the *Journal of the American Dietetic Association*. But the researchers also noted that most of the caregivers with "picky" children offered them a new food no more than three to five times before deciding the child disliked it—not nearly enough, according to Birch. A 2016 study published in the journal *Eating Behaviors* found that 39 percent of kids aged three to eleven were identified as picky eaters at some point in their childhood. The differences in these numbers are difficult to parse out, because although many scholars agree with Birch that neophobia is a normal stage of development, others use the term interchangeably with "picky," "fussy," or "selective" eating, while still others define it as an actual phobia and use the word to describe the most extreme cases rather than a baseline. But the bigger problem may be that the researchers conducting these studies usually ask parents or other caregivers to assess a child's eating habits. How often do kids present with innate and rigid food preferences—and how often do their parents just misperceive them this way?

"For me, the big change, the big epiphany moment was when I realized, this wasn't anything my son was doing," says Skye Van Zetten, a married mother of two in Ontario, Canada. "It was me. I'm screwing this up. I'm ruining this kid's relationship with food through my fear."

Eating started out just fine for Skye's son, TJ, and his twin sister, Dawne. "In fact, when I first gave them cereal at five months, TJ was all over it and my daughter was more like, 'Hmm, I don't know about this,'" Skye recalls. But when the twins were seven months old, TJ choked on a small bite of pear. The ordeal was over in seconds; Skye, who is a trained lifeguard and CPR-certified, knew exactly what to do. "I yanked him out of the high chair, smacked him on the back, and out came the pear," she says. "And it was just like, 'Okay, the pear is out, good, let's eat more pear.' It was terrifying, but he seemed fine."

Skye didn't think about the choking incident again until TJ was two. He choked on some crackers and, again, recovered well. But afterward he wouldn't touch those crackers. The response of Erik, Skye's husband, was, "Great, now there's something else he won't eat!" And that's when Skye realized, "You know what? This kid really does not eat a lot of food." At age four, TJ choked again, this time on a raisin. He dropped more foods and began refusing to come to the table for family dinners. When Skye insisted, he hid under the table or behind the sofa, wide-eyed and terrified. The list of acceptable foods continued to dwindle, until TJ was eating just eleven things: chocolate milk, juice, ice cream, pizza, strawberries, peanut butter, bread, and four types of crackers.

"That was enough for him to go on," says Skye. "But it wasn't enough for me."

Everyone told Skye: "Picky kids will eat when they're hungry enough." So she and Erik decided to get tough. On the first night, Skye made chicken, mashed potatoes, and corn with diced carrots for dinner. TJ refused to eat a single bite. The next day, she served it again and he refused it again. Skye was alarmed but also convinced she was being a "good parent" by not giving in to TJ's resistance. She did allow chocolate milk at the table, but TJ, overwhelmed, drank only a few sips. Finally, on the fourth day of TJ's hunger strike, Skye was done. "Fighting over days-old, possibly spoiled chicken seemed pointless," she says. "I fed my son."

Skye reached out to an occupational therapist, who diagnosed TJ as "an enigma." During the evaluation they played games together, such as pretending to brush their teeth with carrots, but TJ still refused to eat. A pediatric mental health clinic flagged him as a trauma case because of his history of choking, but couldn't make any progress with his eating either. Then Skye read about a clinic in Seattle that prioritized "food exposure" over actual eating, so she served pasta for dinner but told TJ he didn't have to eat anything. "He was relieved and confused." Skye and Eric threw all the normal rules about table manners out the window and encouraged the kids to play with their food. "We all sat there flinging macaroni noodles at each other," she says. "That was the first time I could remember the whole family enjoying food together." But still, TJ stuck to his safe foods.

Then in 2012, Skye began blogging about her experience and found her way to Ellyn Satter, the registered dietitian and family therapist who pioneered the Division of Responsibility model of feeding that we found so helpful in teaching Violet to eat. The key tenets of Satter's philosophy resonated deeply with Skye as well. She already sensed that force-feeding tactics were only reinforcing

TJ's fear around food—but now she saw that so was the "he'll eat when he's hungry" approach, if she didn't combine it with offering the few foods he *did* feel safe around.

On the first night of their new order, Skye put a bowl of TJ's favorite crackers on the table alongside the family's dinner of macaroni with marinara sauce, peas, and carrots. TJ ignored the other food, ate four crackers, and declared himself full. Skye went to the bathroom and cried. But she didn't ask TJ to eat any more. Every day, she put a mix of foods on the table, including one or two of his staples, and then left the rest up to him. Every day she saw him start to slowly unwind, and start to eat a little bit more. Some days, he ate a little less again, as if checking to see whether she'd really let him. But a week later, he surprised everyone by grabbing a piece of bread. He took one bite, then put it down. This time, Skye thought to ask why. "It's too hard," he said. So she got him a softer piece and he ate two of them. "It was me that had to do the changing," she says.

Skye thinks that TJ's early choking experiences may have set the stage for his anxiety around food, but she's sure that her corresponding anxiety and high-pressure tactics played a huge role in exacerbating his fears. "We create the neurosis that creates the need for a diagnosis," is how she puts it. And Skye went into parenting primed with her own neuroses around food. "My family tends toward leaner body shapes, but my mom would go on a diet whenever she gained five to ten pounds," she explains. "I remember watching her wrap her midsection in Saran Wrap before going on a walk to 'melt off the pounds.'" Skye thought her mother viewed any extra weight as a sign of weakness, so she kept an eye on the bathroom scale to make sure she wouldn't be viewed as a failure. "I had all these ideas about what's a healthy way to eat, but it was all about weight," she says. For Skye, that meant losing weight. "When I hit a certain number, even if I felt shaky and faint every time I stood up, that was health, in my mind," she says. When TJ's problems began, Skye became just as fixated on his weight, only now it was about whether he could gain enough on his limited diet.

But as she began to trust her son to eat, Skye says her own thinking about how to eat became far less black-and-white as well. "I had this tendency—and I think a lot of us do this—to think low-fat dairy means no dairy, and lean meat means no meat. And while you're at it, cut out gluten, so there go grains. Now we're down to fruits and vegetables, and fruits have too much sugar," she says. "So vegetables are really the only foods that parents feel good about feeding their kids—and kids don't like them!"

Today, Skye reports, "There are too many foods in TJ's diet to count." He's comfortable around many kinds of fruit and even a few vegetables, and will also eat bite-sized pieces of chicken as long as it's been breaded and baked till crispy. "I have no worries that he'll get where he needs to be with eating in his own time," she says. "And what makes me very happy is that practicing the Division of Responsibility approach seems to also be saving Dawne from the fat-phobic, female-shaming attitudes that I grew up with." At age eleven, Dawne scoffs at the ads in beauty magazines and often asks her mother, "Don't these women know they're already beautiful? Why would anyone believe this garbage?"

Skye has chronicled much of their journey on her blog, Mealtime Hostage, and also runs a Facebook group where she daily encounters mothers who have diagnosed their kids as picky eaters because they only want to eat pasta, chicken nuggets, fruit, and milk. "The truth is, that's actually enough for the kid to live on, but the mother is panicked that it's too much sugar and the child will get diabetes," explains Skye. "I can't just say, 'No, that's wrong.' Many mothers hold beliefs about food that are nurtured by fear and founded on the red herring that appearance equates to health.

You have to wade in gently and delicately pick it all apart." From where she's sitting: "The problem really isn't your kid."

But the problem isn't really parents, either. It's what happens when diet culture invades how parents think about food—including the food their children eat. Eva, Kate, and the mothers in Skye's Facebook group are getting their ideas about how they should feed their kids from the wellness-industrial complex I explored in Chapter 2, as well as from the lifestyle blogs, magazine articles, and celebrity cookbooks that provide much of the average parent's education on early childhood development and nutrition. The so-called lifestyle experts who now champion these causes aren't rigorously fact-checking their claims against nutritional guidelines designed for infants and children. In fact, there has long been a lack of federal nutrition guidelines for babies under the age of two, which leaves parents flailing around for guidance on how to best feed their new eaters. They wind up drawing on a mix of popular dieting wisdom and the alternative-food movement, both of which espouse a near religious fervor for the superiority of home-cooked whole foods. And so an army of bloggers, food writers, and celebrity moms celebrate a kind of continuum of "clean" eating that starts with breast-feeding and continues on to a diet high in protein and vegetables, but low in sugar, dairy, gluten, and just about everything else.

SuperHealthyKids.com is a wellness brand that has built a substantial following (including more than three million Facebook fans) by offering "clean" recipes that they claim kids will actually eat, such as "Rainbow Buddha Bowls" and "Sweet Spinach Muffins." They also regularly post advice with headlines like "5 Reasons to Serve Veggies for Breakfast" and "The 5 Biggest Family Feeding Mistakes You Made in 2016"; the latter list of offenses includes not making a detailed meal plan every week to ensure good health at every eating opportunity, and not taking the time to chop vegetables or cook whole grains in the morning, a failure that leaves you vulnerable to falling back on processed-food shortcuts when you cook dinner. More than 113,000 people follow Jenna Rammell on Instagram, where she challenges them to "Sugarless Holidays" and shows off her kids' lunches, brimming with vegetables and styled into chic stainless steel bento boxes that retail for $40 to $50 each. Catherine McCord, author of two cookbooks and the popular Weelicious blog, offers up recipes for gluten-free black-bean brownies and coconut chia-seed "breakfast pudding," though her 167,000 Instagram followers presumably also appreciate the occasional shot of sugary baked goods, accompanied by winking captions like "weekend balance: donuts then smoothies." Meanwhile, beautifully photographed cookbooks with titles like *Real Baby Food*, *Little Foodie*, *Whole Food Baby*, and *Smart Bites for Baby* promise to help parents train their little eaters to love kale, daikon radish, tofu, and yes, sushi.

Much of this philosophy is rooted in the nutrition ideals of Kate's childhood, all grown up into a hipper, Brooklyn-fied version of the same sprouted wheat bread and almond butter that her mother served, a whole-grain island floating in the 1980s sea of Wonder Bread and Skippy. The mainstreaming of socially conscious, environmentally friendly health food has had plenty of benefits, of course. Locally and regionally grown food is now a $6.1 billion market, according to the USDA, with 150,000 farmers and ranchers selling directly to consumers. And grocery stores no longer think it's weird if you bring your own reusable shopping bags.

But these changes in food culture have also made eating an increasingly anxious and thought-provoking activity. And parents are on the front lines. They have to reconcile these new standards and fears with the reality of their kids' perfectly normal preferences and developmental trajectories, and figure out what to do with a two-year-old who snubs sushi or a preschooler who is disgusted by daikon. The upshot is that we're more alarmed about picky eating than ever before, while also having raised the bar on how kids are supposed to eat. We end up throwing darts: We fight to keep breast-feeding going as long as possible even when it has stopped being what's "best" for us (and maybe for our child, too). We insist on three more bites of broccoli or two more sips of kale smoothie. But then we also give in to the demands for Honey Nut Cheerios and mac and cheese. Meals become a hodgepodge of perfectionism and permission, structure and rebellion. And kids who might otherwise have sailed through a few years of toddler neophobia absorb their parents' anxieties and become anxious themselves. Food becomes a power struggle as kids become increasingly rigid in their own eating preferences. We impart confusing messages to our children about good foods and bad foods because we are so confused ourselves. And our confusion is not just about what they're supposed to be eating, but also about how best to feed ourselves and about how much guilt to feel over our own indulgent falls from nutritional grace.

"I want Annika to see food as fuel, as healing," says Eva. "I want her to enjoy food. I want her not to think about it." I am struck by the disconnectedness of those twin wishes; but for Eva, the combination makes sense. Thinking about food, trying so hard to get it right, has robbed eating of all enjoyment. The only way Eva can imagine Annika feeling good about food is if she can just not care quite so much about it. But Eva's not sure how to get to such a point, when she's never been there herself.

Four

Fear of Food

It's lunchtime on a January Saturday in Connecticut, and Marisa, a high school teacher, is making peanut-butter-and-jelly sandwiches for her kids. Two-year-old Lisa and three-year-old Jonathan are exuberant eaters; they wolf down their sandwiches and several clementines in between climbing on the table and chatting enthusiastically with their parents and with me.

"Do you love oranges?" Jonathan asks me. "You do? Do you love chocolate milk? Do you love everything?" I agree that I do, in fact, love everything.

"So do I," he says, satisfied. "I love everything and my dad loves everything and my mom loves everything!"

What Jonathan hasn't noticed, because he is not yet four years old, is that his mother isn't eating lunch herself. She doesn't even sit down at the table with us. Instead, she stands a few feet away in a corner of the kitchen in their split-level ranch, grazing on

Doritos from an open bag. As her husband, Paul, unpacks groceries, she also grabs a croissant from the plastic carton he pulls out
and eats it quickly, ripping the flaky dough into small pieces.
Marisa, thirty-five, doesn't eat peanut butter and jelly. Nor does she
eat clementines. She doesn't love everything. In fact, the list of foods
that she loves is very short: chips, french fries, white rice, white
bagels, pizza, apples, bananas, dry cereal, toast, and a few other
forms of bread.

Marisa, who asked me to change names for her and her family,
has been eating this way for as long as she can remember. She was
raised by a young single mother who, she says, was "pretty abu-
sive, physically and mentally." Most of the abuse happened when
her mother thought Marisa and her sister weren't listening or were
talking back, but her uncle also recalls watching Marisa being
spoon-fed as a baby. "He used to tell me that when my mom fed
me, she would shove the spoon so far down my throat that I would
gag or choke," Marisa says. She doesn't remember that, but she does
have what she describes as an intense fear of gagging or vomiting.
"Anything unfamiliar makes me gag and therefore that means it's
unsafe," she explains. "All food smells amazing to me, but as soon
as I put the fork in my mouth, it's a different story. I get scared
just knowing I'm going to taste something that has a flavor or texture I've never had before."

Although she has never been formally diagnosed, Marisa
believes that she has a condition known now as avoidant-restrictive
food intake disorder. ARFID (as it is commonly known) is our new-
est and least understood eating disorder; the American Psychiatric Association added it to the fifth edition of the *Diagnostic and
Statistical Manual of Mental Disorders* when it was published in 2013.
ARFID is an attempt to replace a long list of terms, such as "selective eating disorder," "food neophobia," "oral aversion," "severe

picky eating," and "chronic food refusal," most of which sort of
sound like medical diagnoses—and often are even used that way—
but aren't listed anywhere official. Earlier editions of the *DSM*
grouped all these complaints under the umbrella term "feeding
disorders of infancy and early childhood." But that rubric was problematic at best. Diagnostic criteria in the fourth edition of the
DSM required that patients be underweight for their age, which
meant that any child maintaining his weight thanks to a feeding
tube or steady diet of PediaSure didn't qualify, no matter how little
he actually ate. As many as 71 percent of children with feeding
problems do maintain a safe body weight this way, according to a
2013 study by researchers at Penn State University's pediatric feeding program. But they still struggle and without a diagnosis, they
often weren't eligible for therapy and other services. The other big
problem with "feeding disorders of infancy and early childhood"
is that, as Marisa knows, such disorders can last well past the first
years of life. The *DSM-IV* didn't apply the diagnosis to people
older than six; anyone with restrictive food intake in later childhood, adolescence, or adulthood was assumed to have anorexia nervosa or ED-NOS (eating disorder not otherwise specified). But
refusing to eat because you're afraid of food is something quite different from refusing to eat because you're afraid of being fat.

After several years of debate and many position papers, the
expert panel of researchers and psychiatrists who decide these
things for the *DSM* landed on ARFID, which the *DSM-V* defines
as "an eating or feeding disturbance (e.g. apparent lack of interest
in eating or food; avoidance based on the sensory characteristics of
food; concern about aversive consequences of eating) as manifested
by persistent failure to meet appropriate nutritional and/or energy
needs." In order to be diagnosed with ARFID, a patient needs to
meet one of four criteria: they must have lost a significant amount

of weight, developed a serious nutritional deficiency, become dependent on some sort of supplemental feeding (whether through a tube or with oral nutritional supplements), or be experiencing a "marked interference in psychosocial functioning." The *DSM* editors kept that last criterion deliberately open-ended, to allow the diagnosis to capture many of the cases that fell through the cracks of the more rigid *DSM-IV* criteria and thus went untreated. You can now have ARFID and be obese, for example. And the disorder can develop at any age. If she hadn't been born mere months after the *DSM-V*'s publication (that is, before the new diagnostic codes had permeated the medical community), Violet would have been diagnosed with ARFID when she became feeding-tube dependent as a newborn. Adults like Marisa can qualify for treatment (and insurance coverage for that treatment) as well, even if they aren't losing weight or relying on feeding tubes, if they can demonstrate that a diet of french fries and white bagels is taking a toll on their "psychosocial functioning."

One new catch to the criteria: a doctor can't diagnose ARFID if you have another medical or mental health disorder that explains your eating struggles—say, obsessive compulsive disorder or autism—unless the severity of your food issues exceeds what's typical for your primary diagnosis. Some feeding therapists and researchers balk at this technicality, because several studies suggest that ARFID often coexists with anxiety, depression, OCD, and autism spectrum disorders.

How these conditions relate to one another is not yet known, notes Nancy Zucker, Ph.D., the director of the Duke Center for Eating Disorders, which maintains large databases where adults with ARFID, as well as parents of selective eaters, report on their experiences. Zucker also treats ARFID patients through various inpatient and outpatient programs. Broadly speaking, she tends to put ARFID patients into three categories: people who aren't eating enough food; people who don't eat enough different kinds of food; and folks who experience a traumatic event, like choking, which triggers "an abrupt and unprecedented food refusal." But from there, and even within those categories, the lines get blurry, fast.

Altogether, ARFID was found to impact 3.2 percent of Swiss schoolchildren aged eight to thirteen, according to one recent study. Across all ages, prevalence may be closer to 5 percent, estimates Zucker. And when researchers look specifically at populations of eating-disorder patients, they find rates ranging from 14 percent to 22 percent. These are early numbers. As I write this, the disorder is less than five years old, and a search of "avoidant restrictive food intake disorder" in the U.S. National Library of Medicine brings up just fifty-four research papers, while "anorexia nervosa" yields over fifteen thousand, underscoring just how much more science is needed to understand the scale and scope of this newly labeled condition. But if these estimates hold, ARFID may prove to be the most prevalent of all eating disorders—impacting three to five times as many people as anorexia and bulimia, which affect fewer than 1 percent of people in the United States, according to the most recent tallies.

ARFID may also be the trickiest eating disorder to treat. ARFID patients admitted to inpatient eating-disorder programs are often frailer and more underweight than their anorexic counterparts, reported a 2016 literature review by Canadian researchers, published in the journal *Neuropsychiatric Disease and Treatment*. They're also more likely to struggle to gain weight and to require feeding tubes during their hospital stays. Among those who recover, some 38 percent continue to struggle with eating a full year after their initial diagnosis, report researchers from the Cleveland Clinic in a 2015 analysis. And that group includes just the ones who make it through treatment; one 2014 study put the dropout rate for ARFID patients at around two thirds.

At the same time, some therapists argue that the ARFID diagnosis is now too broad and risks pathologizing people who aren't really sick. "Diagnosing variants of normal as being pathological makes the problem seem far worse than it really is," Ellyn Satter argued in her August 2015 newsletter. While I was writing this book, this chapter was the one that piqued most interest at dinner parties because there is something boundlessly fascinating about the notion of an adult who eats like a child. People always want to know: Are they all obese? Or dramatically malnourished? (And even: Do they poop regularly?) A steady diet of pizza and french fries is shocking amid our current vogue for kale and quinoa salads. Those of us in certain elite foodie circles, especially, have absorbed so many messages about the importance of eating our leafy greens and lean protein, it's heresy to suggest that someone could live remotely well never touching anything of the kind.

Underneath those (admittedly judgmental) questions, I've also detected a touch of envy, as if ARFID patients, by eating only the foods they like, are really living out some of our most decadent and forbidden food fantasies. But that's not how it feels to the average ARFID sufferer, or even for people at the subclinical level, who may define themselves as picky or selective eaters. For most of these folks, the forbidden-food list is inverted, with foods that the rest of us define as "good" among their most feared, leaving them only the salty, sugary, carb-heavy options that so many of us only eat when we're "being bad." But is that really proof of an individual psychological crisis—or of a cultural one?

I'll admit that before I met Marisa, I too was expecting to see some signs of her diet in her physical appearance, even though she had already told me that her blood pressure, cholesterol, and other health markers have always been within the normal range. But in person, she seems perfectly healthy, with the kind of shiny dark hair and clear skin that we usually associate with a diet full of blueberries, spinach, and organic salmon. "I do worry that I'm putting myself at risk for diabetes down the road, since I eat so many carbs," she says. "But not enough to do anything about it. Which I guess is kind of sad."

Beyond whatever happened with the spoon when she was a baby, Marisa says, her mother didn't make too big a fuss over her picky eating. She'd cook traditional Spanish dishes for the rest of the family, but make a pot of plain white rice so Marisa could eat with them. There were a few attempts to disguise a bite of steak, say, in rice, and a couple of offers to bribe her with money if she ate a new food, but Marisa wasn't swayed. "My mom did bring it up to doctors and they would say just don't feed her those foods and eventually she'll get hungry enough to eat something else," she recalls. "But I guess my mom never felt comfortable doing that." Marisa's childhood eating habits seemed quirky but not all that bizarre. Many of her friends were picky themselves; later as teenagers, it seemed as if everyone was on one weird fad diet or another, so someone eating only potato chips for lunch didn't really stand out. At age fifteen, Marisa started smoking pot and began to struggle with an alcohol and drug addiction that would last until her early twenties. "I definitely think substance abuse being a priority in my life back then minimized my having to focus on my picky eating," she says.

But Marisa is now twelve years sober. And over the last decade, she has started to feel more anxious about her relationship with food. "I feel self-conscious at weddings or dinner parties, where everyone is supposed to be eating the same thing," she says. When she started her job as a high school teacher, a group of colleagues

invited her to lunch at the local diner, and Marisa found herself
stumbling through an explanation of why she was getting French
toast for lunch. "Most people are curious the first time they see
me eat and then leave me alone after that," she says. But that initial conversation is always nerve-racking for her; she never quite
knows how it will be perceived, or if people are quietly judging
her plate. We just don't expect grown-ups to eat this way.

There have been three times in Marisa's life when she has been
able to make herself eat new foods. The first was when she was
twenty and had signed up for a two-year stint in the Air Force
National Guard. It turned out that her childhood doctors had been
right. When faced with starvation or eating the MREs she was
given during boot camp, Marisa ate. "I can remember one containing ravioli and I just didn't let myself think too much about it. I
was too hungry." Later in her training, cadets were allowed to
choose between their meal trays and the cafeteria cereal bar. With
relief, Marisa switched to eating dry cereal for every meal, even
though walking up there meant passing what everyone called the
snake pit: a table of drill sergeants whose job it was to humiliate the
new recruits. "They would ream me out and make fun of the food
on my tray," she says. "But I just had to take it."

The next time Marisa forced herself to eat was when she was
incarcerated for three months at age twenty-three, for possession
of heroin and cocaine (her record has since been expunged). "In
jail, they don't feed you what you want," she says. "I had to learn
how to manipulate my way through and trade my food for other
people's bread. But sometimes that didn't work." One day, starving yet unable to eat anything on her tray, Marisa stole an apple
off a guard's desk. Another day she tried a small bite of Cup Noodles and, to her surprise, instantly liked it. After that, she survived
by buying ramen noodles and honey buns from the commissary. It
was almost all she ate for the rest of her sentence.

Marisa learned how to eat muffins in boot camp, and to drink instant coffee in jail, and both habits have stuck. But none of the other foods she forced herself to try have remained in rotation, and she says a real downside of both experiences was that afterward she overate all her favorite foods to compensate for the months of repression. "I felt like I had to make up for everything that I'd lost," she explains. "I'm not a heavy girl by any means, but I was for a little while after both of those places." To keep her weight in check these days, Marisa almost always skips lunch. She worries that between breakfast and dinner, she's already eating more calories than she should. But this is where I can see how ARFID differs from anorexia, because Marisa doesn't revel in the restriction. It feels like a necessary evil, and she often finds herself snacking through the afternoon to keep hunger at bay. "I have to have something in my stomach or I'll feel sick."

Marisa's third experience with trying new foods is also the only time she can recall trying to change her eating habits on her own terms. She decided to make the effort shortly before becoming pregnant with Jonathan because she was worried that her restricted diet would be dangerous for a baby. But she didn't know how to go about it; she wasn't sure if this was a job for a doctor or a therapist. So she did some Googling and wound up with a hypnotist, who claimed he could mesmerize her into trying new foods without her usual stress or anxiety. Marisa saw him once a week for three months. But she's not sure it worked. "I don't think I was actually being hypnotized because I never felt that different during the sessions, but I do think it helped to be held accountable," she says. "I had to set a goal at every session for what I was going to try and also report on how I'd done the week before."

Paul was ecstatic—because in what is perhaps the most bizarre detail of Marisa's story, she's married to a professional chef. Paul grew up in a big, food-loving family, worked at a Hudson Valley farmers market for years, and trained at the prestigious Culinary Institute of America in Hyde Park, New York. He's worked in restaurant kitchens all over New York and Connecticut and now cooks for a residence house at an elite university because the hours and lifestyle are more compatible with family life. "I fell in love with the discipline of cooking; there are all these unsaid rules in a restaurant kitchen," he says. "And I love the rush and intensity of it." But he also loves the casualness of cooking at home, where he can cut corners when he feels like it by using store-bought ingredients, but also be creative and come up with different recipes to try on the kids. Jonathan and Lisa love their dad's chili, as well as his red-curry chicken, made with coconut and quinoa. "I try to expose them to lots of different foods," Paul says. This is partly because, as the food-loving parent, he feels responsible, but also just because it's fun for him. "I want them to be my little foodie buddies. But I don't ever force them to try something. I try to remember that sometimes, they're gonna eat like kids."

Each week, after Marisa's hypnosis session, Paul would set about making the best version he could of whichever food she had decided to try. He breaded chicken cutlets one week and salmon fillets another, trying to diminish their meaty textures and bring them closer to the soft, carb-y foods she felt comfortable eating. He tossed mixed greens with sliced strawberries and his own raspberry vinaigrette. Marisa also tried blueberries, pineapples, green beans, carrots, and cauliflower under his watch.

It was a learning experience for both of them. Marisa cried every time she tried a new food. "It was almost easier in prison or boot camp where I had no choice," she says. And Paul learned that he couldn't hype the meal or even talk about it the way he would with anyone else he cooked for. "I couldn't draw attention to it by asking, 'Did you like it?'" he says. "I just had to serve the food and not say anything, just let her do it." Even the most benign comment could throw off Marisa's nerve; if she started actively thinking about the flavors and textures in her mouth, she would get stuck and have to stop. "It was hard when I didn't know all of the underlying rules. I had to learn I couldn't take it personally if she didn't like something," Paul explains. "At the same time, it was so awesome to see her walking through those fears. People just don't understand how hard that is."

Marisa can remember that she really liked Paul's salmon, baked with a coating of panko bread crumbs. She had to eat it very slowly, taking tiny bites so she could be sure each bite would taste the same and offer the same mix of breaded texture. If any mouthful tasted different from the one before, she needed to spit it out. "But as long as there was consistency, it was great," she says. Still, she hasn't eaten salmon again since she stopped seeing the hypnotist. She's kept pineapple, as long as she doesn't take too big a bite, and a few other fruits. But the only food she had a truly positive reaction to was pizza. "That was like fireworks going off!" Paul says. Marisa was surprised by how much she loved pizza. She'd never eaten cheese or sauce in any other form—but the warm dough felt instantly comforting. "Nothing else tasted as good," she says. "It made me mad, that nothing healthier stuck. But nothing else felt as safe."

An intense preference for foods that "feel safe" seems to be one unifying characteristic of ARFID patients, and perhaps all types of picky eaters, regardless of age. Marisa is a member of the Picky Eaters Association's Facebook page, a closed group with slightly more than four thousand members, which describes itself as "a safe place. Safer than the house you grew up in," because it's committed to letting picky eaters air their grievances about food. "NO ONE will make you feel badly for what you DO NOT PUT IN YOUR MOUTH," reads the group's mission statement. The page's cover photo features a collage of french fries, pizza, and grilled cheese, the unofficial staples of an ARFID diet, and many posters defend the nutritional merits of living exclusively on those foods, usually through tongue-in-cheek memes: "Usain Bolt ate nothing but chicken nuggets while he was in Beijing because that was the 'only thing he knew' and he won 3 gold medals," reads one post from January 27, 2017. Group members also regularly post rapturous photos and videos of french fries being drenched in cheese; for a group that claims to hate so many foods, "food porn" is surprisingly prevalent. But whenever a new food image goes up, I see just as many commenters chiming in to say how grossed out they are as those wishing they could eat it for dinner. In some ways, the posts are reminiscent of the troubling "pro-ana" trend found in many dark corners of social media and on weight-loss blogs, where people (usually women and very often, teenage girls) post images of extremely skinny models along with tips on how to restrict your eating as much as possible. Pro-ana content has been banned on Instagram and other sites in recent years because it glorifies eating disorders. Posts on the Picky Eaters Association's page don't always glorify ARFID; quite often they're more about commiserating on the struggle. But there is a theme of self-acceptance that borders on social rebellion. After all, there's no question that someone consistently eating a very small number of foods may be at risk for malnourishment and other health problems. But in this space, such discussions are verboten. Everyone is allowed to be who they are—and eat how they like.

Still, I do detect a kind of repressed longing in many of the discussions. Members often describe feeling sad at holiday meals and other food-based celebrations, or report eating alone in their cars so their co-workers can't judge yet another french fry lunch. They trade tips on how to discreetly spit out offending bites of food at dinner parties, and they joke (but it's not really a joke) about meals that make them vomit. Another ongoing theme is the challenge of losing weight when salad triggers a gag reflex and refined carbohydrates make up the bulk of your meals. "How do I clean up my diet and live a healthy(er) lifestyle when I find all fruits and vegetables repulsive?" asked someone else in one post. This sparked a twenty-five-comment thread of folks debating which kind of blender could most effectively remove the offending pulpy texture of fruits and vegetables enough to make them palatable in soup or smoothie form.

And other members purport to defend such choices, yet inadvertently shame them at the same time. Although the *DSM-V* specifies that ARFID patients don't struggle with the same body image anxieties as people with other kinds of eating disorders, that doesn't mean they're immune. "Many picky eaters eat what they consider, at an intellectual level, to be an unhealthy diet," notes Jane Kauer, an anthropologist at the University of Pennsylvania who surveyed 489 adults about their eating preferences, then did in-depth interviews with 40, for a study published in the journal *Appetite* in 2015. "And most of them get quite a bit of flak from people around them for not eating so-called good foods like fruits and vegetables." The combination can lead to eating patterns that don't look all that different from the behavior of someone with anorexia or another type of eating disorder that is more directly rooted in body shame. In response to a post asking what a "healthy" picky eater's diet could look like, one commenter responded, "Most of my safe foods are 100% junk!" Her solution to that conundrum sounds like classic crash-diet advice: "Starve until your cheat day!"

One thing the Picky Eaters Association members don't talk much about is why their eating habits are so rigid. In part that's because this is supposed to be a place where they don't have to explain or justify their choices. But it's also because the biggest mystery about ARFID—and, indeed, all forms of picky eating—is why it develops in the first place. There are numerous hypotheses: "Is it kids who had bad reflux as babies, so you wonder about early aversive conditioning? Or kids who are highly sensitive to taste, texture, and smell? Or a segment of the population that is just innately more avoidant of food and that was somehow protective at one point?" asks Nancy Zucker of the Duke Center for Eating Disorders. "All of these explanations are possible." The limited research on ARFID has begun to poke into all of them, but nobody has landed on any conclusive answer. As Zucker puts it, somewhat grimly: "We don't know anything."

Lacking other explanations, most picky eaters believe their preferences are an innate part of who they are. "One of the most fascinating things about our research was that all of the people we interviewed felt very strongly that no, this didn't come from anywhere," Kauer says. Her study found a slight correlation with obsessive-compulsive behaviors and symptoms of depression, but nothing definitive. And subjects who identified as picky eaters were no more likely to report an early traumatic experience with food (such as choking) than were the non–picky eaters interviewed. "We also saw no evidence that these people were hyper-responders to other sorts of stimuli. They didn't have sensory overload going on," Kauer explains. "There was just a soul-deep feeling that 'Look, this is who I am, and this is how I feel about myself.'" Yet she acknowledges that there is no research to support the idea that picky eating is an inborn preference.

In the 1990s, Linda Bartoshuk, a psychologist then at Yale Uni-

versity, coined the term "supertaster" to describe people who have
more intense responses to certain flavors, especially bitter ones. Her
research has shown that supertasters make up around 25 percent
of the population and that these folks have more taste buds on their
tongues than the rest of us, which suggests that supertasting is
genetic in origin. Perhaps, in times of food scarcity, supertasting
would have helped people to avoid eating the wrong berries. But
even then, it's hard to tease out how these preferences are driven
by biology and how they're shaped by environmental factors. A
large body of research on supertasters has failed to show that they
are invariably pickier eaters. Some seem to have more refined palates, but that might translate to sophisticated tastes in chocolates
or wines, for example. We may have certain underlying genetic
preferences for particular foods or flavors, but how those preferences manifest depends very much on how we interact with our
food environment. "Historically, we've had many cultures where
people ate a very limited variety of foods," notes Katja Rowell, a
family physician who specializes in feeding relationships and coauthored the book *Helping Your Child with Extreme Picky Eating*.
"The Inuits, for example, historically ate less than twenty foods
most of the time. That was culturally normal for them, but might
be the cutoff point for an eating disorder today." A more familiar example might be the meat-and-potatoes diet that many
Americans have eaten for generations. "Was a middle-aged factory worker in the 1950s considered picky because that's all he
ate?" asks Rowell. "No, because that's all food was then."

Everyone whom Kauer interviewed told stories about being
teased, bothered, and even harassed for their food preferences,
which bolsters the theory that whatever preferences we're born
with are further defined by our experiences. "Picky eaters very
much have a sense that they're under assault around food," she says.

"And every time they're in a situation with a new or not-preferred food, it may feel terrifying." But while several studies, hers among them, support the notion that picky eaters are more likely to have experienced pressure from parents and others, nobody has been able to establish whether high-pressure meals *cause* picky eating. It could be the other way around, that a parent faced with a picky kid is more likely to put on the pressure. "We're talking about really complex behaviors that have developed for reasons we don't understand," Kauer says.

Marisa says she used to wonder whether her picky eating stemmed from a need to find a sort of order within a chaotic childhood. "I used to think I needed to control something in my life and that's all I could control," she says. "And then drugs and alcohol became my escape because I didn't know how to deal with life as it happened." I point out that the two issues could correlate; escaping with drugs was one coping strategy, while restricting her food intake was another. Marisa agrees with my theory, but says that once she found the Picky Eaters Association, her thinking changed. "I highly doubt that the needing-control thing would be the case for everyone who has this," she says. "Part of me thinks it's something way more powerful than I could even understand, happening in my brain."

Another member of the Picky Eaters Association, named Ben, resists the idea that his eating preferences need to have a medical or psychological explanation at all. "I don't think of myself as having an eating disorder, because I do enjoy eating—I just don't enjoy eating a lot of the things that most other people do," says the thirty-nine-year-old stay-at-home dad in Kents Store, Virginia (who also asks to use his first name because he worries about judgmental attitudes toward his eating habits). But Ben also describes himself as "very cautious and risk-averse," and notes that he's been treated for anxiety and depression in the past, so he can understand why some picky eaters identify with a disorder like ARFID. "There seems to be a sort of 80/20 break in the group between people like me, and then a smaller subset of people who don't like to eat at all. It's difficult to understand the position of those who have a real [psychological] condition, though I do respect them."

There aren't many obvious differences between Marisa, who does identify her eating as a disorder, and Ben, who doesn't. He has the same kind of "kids' menu" tastes: lots of pizza and french fries, and he'll also eat some foods, such as hot dogs and very well-done cheeseburgers, that many more selective eaters won't touch. But Ben doesn't like fish, spicy foods, or other strong flavors, like coffee, and suspects he might be one of Bartoshuk's supertasters. He also avoids most cooked vegetables, with the exception of potatoes and corn. "I like foods that are bready, simple, and have no extreme taste," he says. The other common denominator in everything he likes best is that he orders it at a restaurant or gets it out of a package. "I don't like to cook," he says. "And I don't enjoy eating things that have been homemade by most other people, especially if I don't know them well."

I'm tempted to trace Ben's aversion to home cooking back to the dinners of his childhood, where he says he was often forced to sit at the table from five p.m. until bedtime, when his mother would finally relent and let him eat a bowl of cereal instead of his dinner. "I was a good kid and I never got in trouble for anything else, but this was an issue from the time I was four or five years old until I was a teenager," he recalls. But he doesn't see those fraught family dinners as the cause of his eating preferences. "I was going to eat what I was going to eat, either way," he says. "It was just more difficult because my parents were so rigid. It led to a lot of anxiety and fights."

In fact, Ben still feels anxious eating dinner at his parents' house. "I'm thirty-nine, they're sixty-four, and they still don't accept that I don't eat certain foods," he says. "We go there two or three times a year and my mom will inevitably make something she knows I won't eat and say, 'But you need to eat this; it's healthy!'" Ben also dreads dinner parties, where the meal is likely to be home-made and his eating preferences can lead to awkward conversations. "I don't ever want to offend anyone, but there are a lot of foods that turn my stomach," he says. "And I'm just not going to eat something I don't want." Ben says the main appeal of packaged foods is their consistency. He likes knowing exactly how a food is going to look and taste, from beginning to end. He's very particular about his brands and will only eat name-brand General Mills' Honey Nut Cheerios, for example, and only the Target store brand's generic version of Honey Bunches of Oats. It takes Ben a while to warm up to new foods, but once he does, he commits. His wife once suggested that he switch up his order at the Olive Garden from pasta with plain tomato sauce to a dish where the noodles were tossed with olive oil and fresh tomatoes. Ben took the plunge, liked it, and then ordered nothing else at Olive Garden for several years, until the dish was taken off the menu. "I've had that happen quite a few times, where a restaurant or a food brand will stop making the thing I really like," he says. "It's so frustrating, especially when I can't find a good substitute that I like as much."

For the most part, Ben views the challenges of his picky eating as annoyances; his "psychosocial functioning" seems just fine. But he also acknowledges that his diet may have taken a toll on his health, as he was diagnosed with type 2 diabetes nine years ago. Suddenly all the sugary, carb-heavy foods that he loved were off-limits. "It was a little bit scary, honestly," he says. Ben joined a diabetes support group on Facebook and started researching what he could and couldn't eat. "It took me a long, long time before I started to find my own way into trying to eat healthy." These days he tracks everything he eats and is fastidious about portion control, since he's found he can more or less eat the foods he craves as long as he doesn't have too much. Breakfast is usually cold cereal. Lunch is a turkey sandwich on white bread with cheddar cheese and Miracle Whip. If it's cold out, he adds a can of soup— always Progresso's Potato and Bacon or Campbell's Homestyle Harvest Tomato and Basil. Dinner will be pasta, a frozen pizza, or turkey breast with mashed potatoes. In between, he snacks on yogurt or baby carrots. It's all a little bit boring and Ben misses the chips, doughnuts, and french fries that used to make up a bigger part of his diet. "But nobody can eat like that forever and get away with it."

Interestingly, learning about concepts like supertasting and ARFID often seems to bolster the belief of so many picky eaters that they were "born this way." After she learned that ARFID was an official mental disorder, Marisa started to think of her food choices as less of a coping mechanism and more of a disease. It's an understandable shift; if a coping strategy is something we've decided to implement, it's also something we can decide to stop because it no longer serves us. But a disease or disorder feels less optional. That isn't necessarily true; as Marisa knows firsthand from her struggle with addiction, mental health issues can be treated and managed, if not ever fully cured. But she also knows how hard that is, how slow and grinding the progress can feel. When ARFID stops being a personal choice and instead becomes a disorder, fixing it feels like something she has to outsource to experts rather than exploring what she could do to help herself. Which is not to say she's given up. Rather, she, Ben, and many members of the Picky Eaters Association focus on developing different skills. They know how to give short explanations of their conditions, how to graciously deflect attention from their food choices, and how to plan ahead to make sure they'll have something to eat most places they go. They know how to navigate the world as picky eaters, despite how often the world seems to be programmed against them.

Marisa takes Lisa up for her nap while Jonathan finishes his lunch and Paul begins mixing up meatloaf for a dinner that only he and the kids will eat. Jonathan is momentarily excited to help stir things, then wanders off to play with Legos. When Marisa comes back from putting Lisa down, she sprawls on the living room floor to join him. "I just stay out of anything to do with food," she says. "Like if Paul is cooking, I'll usually stay out here. And we don't always eat together. Sometimes I'll just stay in here while they're eating in there."

Marisa says it's not a problem to be around food as long as nobody's expecting her to eat it; it's more that she doesn't want to distract the kids from eating Paul's meals when they see Mommy having white toast and nothing else. When the kids were babies, Paul made all their baby food from scratch, stocking the freezer with creative vegetable purées. Now he packs the daycare lunches in addition to cooking their dinners, except for the odd night when he's caught at work. Then Marisa adds chicken tenders to the tray of french fries she's making for herself. "My first instinct is always to offer them junk food, I guess because I know that's what I want," she says. "It's really hard to remember to offer them something I won't eat." Perhaps as a result of Paul's influence, neither child seems to have inherited her palate. "I'm trying to raise two eating partners who will want to go out and enjoy different restaurants with me," says Paul. The family considers Lisa, in particular, "a little foodie"; they tell me how the night before, she happily ate the spicy Chinese food that Paul ordered. (Marisa had two slices of white toast.) "It will take so much stress off me," Marisa says of a future in which Paul can take the kids out to explore different cuisines and the kinds of five-star restaurants he trained in. "I hate going to those places and just ordering fries. But I also hate that he misses out on those experiences." Paul shrugs off her concerns. "I get to scratch that itch at work; it's one of the good things about me being a chef," he says. But later, he does acknowledge: "The hardest part for me is that I feel like I know what she's missing out on."

And that's one of the biggest sticking points between picky eaters and the rest of us. Marisa doesn't feel as if she's missing out. She enjoys the foods she eats, and if it didn't seem so weird to the rest of us, she'd happily eat that way forever. "I like the textures. I like that it looks plain, that it's all one color," she says. "These foods taste good and they fill me up." She says she does sometimes get into ruts, and is careful to deliberately rotate her meals to avoid that—a dinner or two of pizza must be followed by fries or rice the next night. But if anything, the fact that Marisa eats such a short list of foods, and often in such careful quantities, makes her enjoy what she can have all the more. When I ask how ARFID has affected her lately, she tells me about a batch of fries that sat out long enough to get too cold and soggy for her to eat. "When food is not satisfying to me, I get angry," she says. Of course, most of us don't like cold fries, but for Marisa, the ruined fries were more than a bummer. They inspired rage and also a sort of panic, as she tried to figure out what else there might be in the house that she could eat.

Marisa acknowledges that how she eats is "a huge part of who I am; if I had to give it a percentage, I'd say 50 to 60 percent of me." But as much as she'd like to be able to eat differently, especially in social situations, she doesn't feel that food is in control of her. She worries in an abstract sense about the long-term health impact of her diet, but knows that she's unlikely to overdose on french fries. At this stage in her life, it's unclear whether Marisa would meet any of the diagnostic criteria for ARFID: as an adult in charge of her own decisions, she can live with the interferences in "psycho-social functioning" caused by her eating habits. The ripple effects of her condition are fairly limited, especially with Paul so capable of cooking for himself and the kids. "I feel like I run it. I pick what I want to eat. I guess mentally, I feel like it runs me sometimes, but I don't think about that aspect of it," she says. "I'm in control of what I eat." Still, a few months later, I see a post on the Picky Eaters Association page: "If you could change or get rid of your picky eating, would you?" a member asks. Marisa's posted response: "Yes! Without even thinking twice."

It would be too simple to say that ARFID is purely a product of a judgmental food environment, or that solving it is as easy as being able to live life on your own terms, grilled cheese sandwiches and all. For one thing, self-acceptance around food isn't easy for almost anyone—but especially when the origin story of a person's eating habits is so mysterious. Whatever the initial trigger, for the thousands of people who are admitted to eating-disorder treatment programs each year, as well as the thousands more who aren't able to access help, ARFID can and does spiral beyond something they can manage on their own, or with a supportive family. And when that happens, the path to recovery is far from straightforward.

Jennifer, who asked me to change her name for reasons that will soon become clear, began pretending as a small child that she was too sick to eat. She's been told that as a baby, she did fine with bottles and purées, but began gagging and spitting out food as soon as her mother started serving chunkier baby foods. From that point on, meals were a battleground. Jennifer's mother, whom she describes as "the sweetest woman alive," tried not to press the issue too much. But she also said nothing when Jennifer's father insisted on clean plates. There were nights when he made Jennifer sit at the table for hours until she cleared hers. When the family went out to dinner, her father would grumble about all the restaurants they couldn't go to "because Jennifer won't eat anything." It was easier for Jennifer to say, "I'm fine, I'm not even hungry," and wait out the meal than to order something she couldn't bear to eat. She did the same at family get-togethers: "There were several Christmas dinners where I would say I was sick and go lie down and try to sleep while everybody else ate," she says. And that approach also worked at friends' houses, where Jennifer would also try to be really polite and well behaved before she sat down at an unfamiliar dinner table. "I wanted to stack up as much good about me as I could before they realized I was weird." The truth was, Jennifer was hungry much of the time. But she got used to feeling that way, and she would always choose hunger over trying to eat something that scared her.

Now thirty-nine and an office assistant in a Midwestern suburb, Jennifer still eats the same short list of foods she liked as a kid: peanut butter sandwiches, pizza, grilled cheese, french fries, and what she calls "the mashed potatoes of shame" because they're all she ever eats on Christmas and Thanksgiving. There are a few other safe foods, like the protein shakes she drinks for breakfast most mornings, the English muffins with peanut butter that she packs for lunch, and the cereal—Multi Grain Cheerios, Kix, or Grape Nuts—she has for dinner. A few years ago, she managed to add yogurt to the list, but only vanilla-flavored Dannon Light & Fit Greek Yogurt. "It's thick, so it's not so slimy. Yoplait is a little bit too runny," she explains. "And I can't do chunks of fruit or any-thing else in it." Jennifer eats no meat or fish and very few fruits and vegetables. Unlike many people who identify as intensely picky eaters, she doesn't even like mac and cheese, though she can man-age a few forkfuls if pressed. Trying new foods is a terrifying process. "I really want to eat an orange. Like, so bad," Jennifer says. She loves orange juice. She loves the smell of oranges. But she can't make herself eat one. "Every time I bite into an actual orange, it feels like I'm biting into a baby's finger. There's that squish; it has a sort of skin and then it's all soft and runny inside and I just can't do it."

By seventh grade, Jennifer had stopped eating lunch. Her teach-ers no longer kept tabs on who ate what, so she was free to go out-side instead of dealing with the cafeteria, which was full of food she couldn't stand. She also didn't eat breakfast most days, because even a simple bowl of cereal seemed to upset her stomach first thing in the morning. Which left those tense family dinners as her only eating opportunity of the day. If she had to stay after school for a play rehearsal or other activity that lasted through dinner, she might go to bed without eating at all. And those days felt like a relief—and also a victory. "I never woke up the next morning thinking, 'I have to eat breakfast,'" she says. "That's how much I hate eating. I hate eating in front of other people. I hate the fact that I have to eat, period. To this day, every time I do it, it feels like a personal loss."

This is where the shame of picky eating can start to look a lot like a body image–based eating disorder. Jennifer's childhood inability to eat what she and her family considered to be the right foods has led to a lifelong conviction that she doesn't deserve to eat at all. But prior to the addition of ARFID to the *DSM-V*, many doctors wouldn't have known how to diagnose Jennifer, because despite such frequent fasting, she has always been overweight.

She might have qualified for a diagnosis of "eating disorder not otherwise specified" (ED-NOS), but nobody ever suggested it. When Jennifer was fifteen, her mother tried to discuss her eating issues with a doctor, who responded, "Ma'am, there's no doubt about it. Your daughter is fat. F-A-T fat." He wanted to put Jennifer on a twelve-hundred-calorie-per-day diet—not knowing that most days she barely ate that much anyway—and send her to "fat camp" for the summer. After the appointment, her mother tried to talk cheerfully about the new diet as they drove to the grocery store, but Jennifer was panicking at the thought of being shipped away someplace where people would force her to eat. She told her mother, "If you send me there, I'll never speak to you again." Neither the diet nor fat camp was ever mentioned again. That was the last time Jennifer's parents tried to seek professional help for her.

In college, Jennifer would sometimes go without eating for as long as two days. She didn't get a meal plan because she assumed she wouldn't be able to eat anything in the school cafeteria. She often skipped meals until one of her roommates ordered a midnight pizza or picked up bags of potato chips for the dorm. Then she snacked late into the night. She gained twenty-five pounds her freshman year. "I was so disappointed with myself," she says now. The pattern continued; when Jennifer could avoid eating long enough, she lost some weight. During her junior year of college, when she lived with a thin friend, Jennifer rarely ate at all in their apartment because she was sure her roommate would think, "Well, look how you eat; that's why you're so fat." The next year, when she lived with a roommate who was heavier, she began eating at home again. "I didn't feel so ashamed to eat in front of her," she says. "Of course, I put the weight right back on."

Unlike Marisa's eating habits, Jennifer's appear to have taken a toll on her physical health, along with her mental well-being. She

was diagnosed as prediabetic at age twenty-six, although the diagnosis was later rescinded because starting a course of metformin
made her blood sugar drop too much. She was also found to be
anemic around the same time, though it's unclear whether that
was because of her diet (she eats no meat, spinach, or other iron-
containing foods) or menstruation issues; when Jennifer was in
her late twenties, her period started and didn't stop for six months.
After months of failed interventions, her doctor advised a hyster-
ectomy because it seemed the only sure way to stop the bleeding.
"Now I look at my friends with kids and think, 'I'm never going
to have that,'" she says. "But I also think I dodged a bullet. I was
completely terrified to have children because I had no idea how I
was going to feed them."

When Jennifer was in her early thirties, a new doctor suggested that she might have an eating disorder. (ARFID didn't yet
exist, but some doctors were diagnosing symptoms like Jennifer's
as "selective eating disorder.") "I was telling her how I wanted to
lose weight because of the diabetes risk and we started talking
about diets," Jennifer says. "So then I had to explain that I like
very few things and I don't even like to try a different brand of
the few things I do like." Her doctor was sympathetic, and Jennifer felt temporarily vindicated. She did some research online
and realized she wasn't the only adult with such limited food
preferences. "I thought, 'Okay, it's not just me being difficult,'"
she says. But her doctor didn't offer much in the way of a solution,
leaving Jennifer to double down on her belief that her eating habits
are pretty well fixed. "I like to think I was just born this way,"
she says.

Even if that doctor had known to refer Jennifer to an eating-
disorder center, it's not clear that she would have gotten the help
she needs. Most eating-disorder centers focus on refeeding strate-

gies, treating patients with ARFID much as they do patients with anorexia. Inpatient treatment for anorexia may involve feeding tubes and mandatory mealtimes, while the patient's fears about eating are worked through with a combination of one-on-one cognitive behavioral therapy and group discussions. Once the patient is stable enough to transition to outpatient care, parents are trained to refeed their starving child through carefully structured family meals, in a program known as family-based treatment. FBT can, in some ways, sound antithetical to the Division of Responsibility method created by Ellyn Satter, which we used with Violet. Instead of assigning set jobs to each side of the feeding relationship, FBT works by giving parents total control over their child's eating. The parents decide what, when, and how much, and then—lovingly but unflinchingly—insist on a child's compliance. "You can't use Division of Responsibility in the acute stages of anorexia because kids don't have the capacity to make healthy choices. But this is how we get them back there," says Harriet Brown, a journalist who detailed her own family's experiences with her older daughter's anorexia in her book *Brave Girl Eating*. She acknowledges that some FBT strategies—like sitting at a table with your child while she cries and rages, calmly insisting that she eat—can look harsh to an outsider. But the intention behind them is very different from the kinds of high-pressure family mealtimes experienced by many ARFID patients. The focus on refeeding and FBT makes sense in the acute stages of anorexia, when patients will otherwise starve themselves, quite literally, to death. This is not "clean your plate" as punishment, but as salvation.

The development of FBT in the 1990s also represented a breakthrough in understanding the cognitive inflexibility of the starving brain. The repeated act of not eating reinforces the refusal to eat on a biochemical level. This recognition was an important shift away from earlier thinking about anorexia, which blamed parents (especially mothers) and held that a patient's only hope was to be taken out of the family's food environment completely. "The eating-disorder field maligned parents for so long and we've really needed to atone for it," Zucker notes. There is good evidence that high-pressure family meals or parents who model disordered eating patterns themselves can contribute to a child's dysfunctional relationship with food. But researchers now view anorexia as a condition much like autism, with genetic predispositions that can be triggered by any of a myriad of events. "Often, they don't lose the initial weight on purpose; they start playing sports or get the flu and that experience is reinforcing to them," she explains. "It could be their biology, it could be their sensitivity to how their body feels, or having a driven personality." But interestingly, a history of picky eating preferences is unlikely to be a factor as the data linking early picky eating to eventual anorexia is surprisingly weak. In part this is because there is no standard scientific definition of "picky eating," and researchers generally have to rely on the memories and impressions of patients or their families, which are difficult to vet for accuracy. But when researchers interviewed the mothers of 325 anorexia patients about their daughters' early eating habits, they found no connection between rates of picky eating and the later development of the disease, according to a 2012 study published in the journal *European Eating Disorders Review*. "Anecdotally, my folks who develop anorexia often had the exact opposite experience," says Zucker. "They loved food as kids and are afraid of their emotions, of how much they loved food and the vividness of that experience. It makes them want to shut everything down."

Zucker's theory is that once you get past the initial triggers, all eating and feeding disorders are really about the same thing: a distortion of the hunger and fullness cues that help us understand when and what we need to eat. "Whether they're restricting, throwing up, or having sensory overload, these different disorders are all just about manifesting symptoms that distract people from being comfortable in their own bodies," she says.

But that doesn't mean that all eating disorders will respond to the same type of treatment. A focus on refeeding and FBT may be far less helpful—and even downright damaging—for ARFID patients. Many eating-disorder centers set inpatient refeeding goals at two pounds a week for all underweight patients. "But with ARFID, you're talking about someone who may be small, but they've had stable growth for the past several years. It doesn't make sense to prescribe them three thousand calories a day when we have evidence that pressuring or coercing children around food can lead them to eat even less. Especially if pressured feeding practices are a part of that child's history, then these kinds of therapies are unlikely to help," says Katja Rowell, who travels around the country giving presentations on ARFID to eating-disorder centers. "Nobody defines what the *DSM* means by 'failure to maintain adequate growth' or 'weight loss.' But we know that children being small for their age is the number one reason that parents get into pressure with feeding." Which again brings up questions about the role of parental feeding style: Are so many patients with ARFID underweight because they've been restricting for years? Or are they just naturally smaller, and did the problems with food develop in response to parents made anxious by growth charts?

Either way, if parental pressure has played any role in the development of a given patient's ARFID, it's hard to see how FBT, with its focus on parent-led mealtimes, will help. Even with anorexia, Brown notes, "ideally FBT is done with the family of origin, but that doesn't always work." In cases like Jennifer's, FBT would drop the patient right back at the table where so much of her early trauma around food took place—only this time, with a therapist agreeing that she has to eat whatever her parents say. "If you have a twenty-two-year-old who has struggled to eat since they were first introduced to solids, I don't understand how you can treat them without asking questions about their early mealtime experiences," says Rowell. "Someone who has dealt with this since they were six months old shouldn't be treated the same way as a fifteen-year-old who has gone on her first diet, lost weight too rapidly, and now needs refeeding."

The other strategy most often used with ARFID patients is an exposure-based therapy that uses a few techniques reminiscent of the approach employed by the behavioral pediatric feeding programs I explored during our quest to help Violet. At the University of Pennsylvania, Hana Zickgraf, a co-author of Jane Kauer's and a doctoral student in clinical psychology, counsels ARFID patients through the School of Medicine's Child & Adolescent OCD, Tic, Trich & Anxiety Group. "The most successful patients we've had were willing to eat a small amount of a new food every day, and slowly increased the portion until they were ready to keep it in their diet," she explains. "At first it's slow and agonizing and they accomplish one food at a time. But eventually they start to generalize; if they've gotten used to shrimp, then other kinds of shellfish or even regular fish don't provoke as much of a neophobic response." Zickgraf teaches her patients how to minimize their gag response through specific chewing techniques, and also by alternating bites of the new food with bites of a safe food to help them wash down each mouthful. She doesn't use "escape extinction" or any of the more controversial techniques used in pediatric feeding programs, although she believes those approaches have their place with younger patients: "You couldn't do that with older children [or adults] or it would really look like force-feeding."

But even in the absence of the more draconian aspects of the behavioral approach, Rowell worries that exposure-based therapy is often too goal-oriented and not sufficiently patient-led. She believes that teenagers and adults with ARFID should be treated in much the same way she works with children, through a combination of Satter's Division of Responsibility to repair toxic family mealtime dynamics, and mindful exposures to new foods. "My approach is, 'Let's just be curious and see what happens with this apple,'" she says. She does experiments with patients in which they might try slicing an apple into thin or thick pieces, and then taking small or big bites to see what feels better. The process mimics the kind of explorations a typically developing baby might do with his first solid foods, in social settings where he can see other people enjoying food. "They have an out every step of the way. We let the process of discovery unfold and let the pace be what it is, rather than setting hard goals like ten new foods in eight weeks. Maybe you can attain that, but the client is gagging and barfing the whole time."

The drawback to Rowell's approach, as with the child-led feeding specialists from Chapter 1, is a lack of clinical data showing efficacy. It's difficult to put her approach into a clinical trial because it's so specific to each patient and therefore difficult to scale up and randomize. But when it comes to ARFID, she's hardly alone on that front. Clinics are beginning to track data, but they mostly focus on rates of weight restoration and program compliance: How quickly did patients get back to a so-called healthy weight, and how did they tolerate the various therapeutic approaches used on them? That's quite different from measuring what Rowell calls "eating competency," where she would ask how much anxiety they feel around food, how confident they are in choosing what to eat, and whether they're starting to derive any comfort or pleasure from mealtimes. "We don't have an evidence-based gold standard treatment yet," she says. "The best any of us can say is that we're trying to be evidence-informed in our approach. So why can't we say that being able to enjoy mealtimes is one of our goals?"

It's a debate that leaves people like Jennifer a bit nowhere—because the goal of enjoying mealtimes feels like an impossible reach. Like Ben, Jennifer still sometimes braces herself for meals with her parents, when comments about her food choices inevitably seem to come up. But unlike Ben and Marisa, she doesn't enjoy most of the meals she eats as an adult, either. Her husband of thirteen years is understanding about her eating; Ryan (whose name I've also changed) likes a pretty limited set of foods himself and will happily eat the same thing over and over, though his palate isn't as restricted as hers. "It's more like he just doesn't care about food," she says. "I have guilt that I'm not a better cook, but whenever I do make something, he always says that he likes it. I think he's too nice to discourage me." The rest of the time, the couple tends to prepare their meals separately, or order out. Ryan often works nights, which means Jennifer is frequently on her own in the evening; dinner then is a bowl of cereal, except when it's an entire carton of cookie-dough ice cream.

Ryan is less patient with Jennifer's obsession over her weight. Over the past four years, she has lost seventy-three pounds by obsessively recording everything she eats, weighing herself daily, and banning several of her favorite foods. She hasn't touched a potato chip in seven years. More recently she gave up ice cream and, this year, soda. "Every year I give up something for New Year's, even though I know I don't have that many things to give up," Jennifer says.

I talk to Jennifer on an evening when Ryan is at work and she's otherwise facing a lonely few hours of TV and cereal. Calls like these are strange for both reporter and source; you quickly forge an intimacy that seems to hang apart from the fact that you've never met in person. We've been talking for nearly two hours when Jennifer abruptly jumps into another story that, at first, seems like a tangent. When she was ten, her brother showed her how to drive the three-wheeler around their parents' property—and Jennifer drove it into a garage wall. She shattered her cheekbone and dislocated her jaw. Her mother rushed her to the hospital, which admitted her for the night. And upset though she was about the accident, Jennifer was more freaked out by that prospect: "I thought, 'They're going to feed me in the morning, and what if they don't have cereal, what if they bring bacon and eggs and I can't touch it?'"

As it turned out, the hospital breakfast was just cereal, and anyway no one forced Jennifer to eat, because it hurt too much to chew with her facial injuries. It wasn't just that nobody forced her to eat; they seemed to understand why she didn't want to eat at that particular moment. She was allowed to feel the way she wanted to, about food and about the whole experience of being in the hospital. That's not how it worked at Jennifer's house, where her father's moods dictated whether the whole family had a good or a bad day. "The only time I got to feel how I felt was when something hurt. And the only time I didn't have to eat, was when something hurt," she explains.

Jennifer had more falls, not all of them accidents. She got concussions. She cut herself. "I didn't really excel at that so much. But what I really enjoyed was punching myself," she says. I can tell that a part of her is relishing telling me this, maybe because she expects me to be shocked, maybe because it just feels safe to tell this kind of thing to a stranger on the phone, thousands of miles away. "I punished myself for being me," Jennifer says.

There are so many ways to read Jennifer's story. Did the picky eating lead to her anxiety, depression, and self-harm? Would she still have ended up here if her parents had known better how to handle their daughter's refusal to eat? Or are her food choices merely symptoms, tiny stars within a larger constellation of mental health issues? It's impossible to know. And that makes stories like hers easy to dismiss, both as something bigger than her food issues, and also as somehow smaller, because it all seems too weird, too sad, to have much to do with the rest of us. But I think it's more useful to think about the similarities between Jennifer's story and our own. Most of us can remember at least one childhood dinner when we were pressured to eat more than we wanted. If we think back, we can recall how dry and tasteless that food became the longer we sat there, how our throat closed up around each bite, how hard it is to eat when you're crying hot, angry tears. And most of us can also remember feeding someone else—a child, a younger sibling, even a picky spouse or friend—and urging them to eat more, have seconds, just try a bite, because we made it, because we think it's good for them, and because we need them to think so too. Food and love are inextricably linked in most families, but so are food and power. And none of us learns to eat without realizing that, somewhere along the way.

A few weeks after we talk, I see a post from Jennifer on the Picky Eaters Association page. "So over the last few months I've just not been interested in food," she writes. "I mean, yeah, I eat but only because I know I should." She goes on to describe her shake-toast-cereal diet and how sick she feels when she tries to eat anything else. "I did ok with the mashed potatoes of shame at Christmas dinner but it's getting to the point that I just really don't want to eat at all." Some of the people who respond clearly think she's exaggerating; they're the Ben-style picky eaters, who can't get enough of their favorites despite the long list of foods they won't eat. But I think of something else Jennifer told me when we were talking about her New Year's resolution diets. "I do wonder," she said. "Is there going to be a point where I say, 'This is the year I'm giving up food altogether'?"

Five

Eating While Black

There is a dog-eared manila folder pinned up with a green thumb-tack in a corner of Sherita Mouzon's cramped kitchen in North Philadelphia. The folder is spread open, poster-style, and all across it, Sherita has taped up photos cut from magazines: A tiny white model with spray-tanned abs. A jar of bone broth protein powder. Another thin white model wearing a sporty black bikini. Grilled chicken kebabs. And the Before and After pictures of someone named "Kelli from Los Angeles" who lost thirty-nine pounds thanks to a special diet supplement regimen and can now fit into her red string bikini. Clipped on top of all the photos is a hand-written list of the supplements that Sherita takes daily: omega-3s, green tea extract, whey protein, and something called pre-probiotics.

Sherita is a very pretty, moderately round, forty-year-old black woman. On this day in April 2017, she has just come back from the gym and is wearing workout pants and black high-top sneakers along with lots of chunky rings and earrings. I know from her Instagram feed that she wears all that jewelry while actually exercising, which puts her in a category of women I have long admired but will never be. She also wears dramatic false eyelashes, and her natural hair is scraped back and styled with a waist-length black ponytail. "Go check out my vision board," she says, waving me into the kitchen for a closer look. "It's helping me stay on track, stay healthy."

"It's . . . a lot of skinny white girls," I say.

Sherita laughs. "I know, girl, but they don't put black beauties like me in the fitness magazines."

Sherita goes to the gym almost every day and has a near religious zeal for healthy eating. Her small kitchen is stuffed with food: loaves of whole-grain bread, boxes of granola bars and organic cereals, and bottles of water are stashed in each cabinet and spilling onto every available surface, along with the array of vitamins and supplements. Her freezer, which she opens proudly, is loaded with bags of frozen broccoli, peas, kale, and collards. Sherita loves to post photos on Instagram of what she's made for dinner—tilapia and broccoli, roast chicken with quinoa, salmon with baby spinach. Every meal is styled carefully on white Styrofoam plates. She doesn't own a full set of real ones.

This is because Sherita and her family live only marginally above the poverty line. Her husband works as a materials handler at a pharmaceutical warehouse; she runs support groups for people struggling with trauma and drug addiction, both experiences that she lived from an early age. "We had no heat. We had no hot water. We were shitting and pissing in buckets," is how Sherita describes her childhood. The tiny row house she lives in now is cramped and run-down, with a backyard choked in weeds and barbed wire fencing, but at least the family can afford to keep the utilities on. Although they have heat and hot water now, Sherita still doesn't take showers. "I just wash up in the sink," she says, patting herself with an imaginary washcloth to demonstrate. "That's what I know."

For most of her life, Sherita has also known hunger. Many of the places she lived lacked a working kitchen. Sherita remembers her mother cooking the occasional dinner of Hamburger Helper or fried eggs and cereal, but mostly, she fended for herself. "Eating healthy wasn't taught to me coming up in poverty," she says, adding dryly: "I was not raised on salmon and brown rice and super-foods." Occasionally, her family would go downtown to walk around at Philadelphia's famous Reading Terminal Market and Sherita would dream about eating all the food she saw there but couldn't buy. "I think I always had a love affair with food because I could never get what I really wanted."

Thirteen percent of Americans live in what academics call food-insecure households, which the USDA defines as those that report some degree of anxiety over food shortages, that face a reduction in "quality, variety or desirability of diet," or that have dealt with multiple instances of "reduced food intake" and "diet disruption." In other words, people who are food insecure aren't able to eat a wide enough variety of nutritious foods, or they aren't eating enough, period.

Our cultural narratives of poverty too often paint it as a problem only of black communities, and I want to be clear about why that's wrong. Chronic hunger exists in every county of every state and among every ethnic group, although most often among families with children. Nineteen percent of Hispanic households and 10 percent of white households face food insecurity. It's happening in cities, in suburbs, in rural communities—and too often, it's discussed least in those places where we don't expect to find it. "I see the other moms getting together for coffee at Starbucks, which I can't do now," a white suburban woman named Amy Bardwell told me when I interviewed her about relying on food pantries while her husband was out of work, for a *Parents* magazine story in 2011. "It can feel surreal because we live in a pretty affluent neighborhood." In fact, a quarter of food-insecure people live in households earning more than 185 percent of the poverty line, meaning they struggle to afford food on incomes that are still too high to qualify them for government assistance.

Nevertheless, in this chapter, I have chosen to focus on the stories of black families because they are still the Americans hit hardest by both hunger and poverty. More than one in five black households qualify as food insecure, the highest rate of any individual ethnic group. The ten U.S. counties with the highest rates of hunger also have populations that are at least 65 percent black. And in 2015, 24 percent of black Americans lived in poverty, compared with just 9 percent of white Americans. So black people are disproportionately affected by this issue. But the real reason I wanted to explore the experience of hunger in poor black communities is that these neighborhoods have been the primary focus of the alternative-food movement's quest to fix hunger in America—and those efforts reveal much about our nation's complicated relationship with race, class, and food.

When I first met her in 2011, Sherita was still using cocaine and struggling to find work that paid better than her last job, as a home health aide. We sat in Sherita's living room while her then five-year-old daughter, Joeanna, watched cartoons and Sherita told me how she worried that her cocaine addiction contributed to the little girl's health problems, which included an early failure-to-thrive diagnosis and some developmental delays. Sherita managed to stay off drugs during her pregnancy, but relapsed hard afterward while in the throes of postpartum depression. Her husband was working long hours and she didn't have much of a support network, since most of her friends were just people she partied with. "You try being stuck at home with this crying-ass, hungry-ass baby, no money to do anything or go anywhere," she told me. "You know how people talk about hearing voices telling them to drown their kids in their bathtubs? Yeah, I heard those too."

Cocaine, which Sherita called "my white god," felt like her only means of escape. "It keeps you up, it makes you sexual," she explained. And it kept her from thinking about food. "When I was high, I didn't care what I ate, or what Joeanna had had to eat that day. I could care less." Of course, the subsequent low—when she couldn't afford her next fix, when she realized that the kitchen was empty and her daughter was yet again crying and hungry—was as terrible as the high was great. When I asked Sherita what she did then, she laughed. "I'll tell you what I'm trying not to do, which is going back to being an escort," she said. "But a child gets tired of hot dogs and fucking Oodles of Noodles all the time. For me, it's like, 'Ooh, look at that old white guy over there . . . I'd probably get a hundred or a buck fifty right there.' And then we can go to Whole Foods as opposed to heating up them hot dogs again."

Society expects a recovering cocaine addict to view prostitution as the means to a drug score. It's a stereotype that Sherita resents, but also sometimes invokes herself when she talks about "the drama" her neighbors get up to on their block. But to consider prostitution as the means for a trip to Whole Foods gets at something else. It shows how Sherita has long wanted a different life, and how limited her options are to get there. She may not have grown up eating "salmon and brown rice and superfoods," but even during that tenuous stage of her recovery, that's the level of health—and wealth—she aspired to for herself and for Joeanna. Because what Sherita hates even more than the escort/addict stereotype is the perception that poor black people don't know or don't care about healthy eating. "People think we're stupid and we don't know that's unhealthy," she told me one afternoon as we walked through the grocery section of her neighborhood dollar store, which is lined with paper cups of ramen noodles and generic canned meats. "We know. Of course I want my baby to eat green smoothies and kale, but that shit is too expensive."

Four years later, the row house is still run-down and the block hasn't gotten any better. But Sherita's husband has kept his job and Joeanna has caught up on her early delays and is now flourishing. Sherita herself has been working more consistently, and is now almost five years sober. Their life feels a little less precarious. Yet the old fears—of empty cupboards, of roach-filled kitchens, of going to stores and not being able to buy anything—have all persisted in ways that are perhaps even more powerful than her cocaine craving. "I have to have food in the house at all times, even though I'm not hungry anymore," Sherita says. Whole Foods is still too expensive, but she loves to go to ShopRite or Trader Joe's by herself, and wander the aisles, soothed by the feeling that she can now buy anything—or at least most of the things—she sees. "I don't need a drug meeting to stay off coke," she says. "I just go to Shop-Rite with money in my pocket. It makes no sense how often I go to ShopRite, but that's my trauma now."

Despite her overflowing kitchen cupboards, the days when Sherita most feels as if she's made it, as if she's broken the cycle of poverty, are the days when she eats as little as she did at the height of her hungry years. "I try for less than a thousand calories a day. Maybe five hundred," she tells me. This isn't out of financial necessity. This is her vision board at work. Now that she no longer considers herself poor, Sherita wants to be skinny. She wants flat abs and a tiny waist. Because she knows it's a privilege to diet.

Sherita wants to "eat clean," which she defines as a Paleo-inspired diet of lean protein and organic vegetables, with a heavy emphasis on alternative-medicine staples like bone broth and omega-3 supplements. She's switched Joeanna's juice and hot dogs to organic brands as well. "Not that Applegate Farms is really healthy, it's still hot dogs, but it's better," she says. "Or it better be, at seven dollars a pack." Sherita's information about nutrition has been gleaned from the fitness magazines she uses to make her vision board, as well as the celebrities and wellness gurus she follows on social media. There are micro-differences, of course: Eva in Los Angeles ate artichokes to promote fertility, and Sherita buys protein powders that promise to aid in her body-sculpting ambitions. But this is still the same modern diet culture we've encountered over and over again. It's the same movement that originated in the patchouli-scented health-food stores of the 1970s and then fully entered the zeitgeist in the early 2000s with the rise of big health brands like Applegate Farms and Whole Foods, as well as celebrity food activists, writers, and chefs like Michael Pollan, Mark Bittman, and Alice Waters, all of whom rose to fame preaching the need to eschew processed foods in favor of locally grown, sustainably farmed fare that "your great-grandmother would recognize."

As part of the quest to fight obesity, save American farming, and teach everyone to appreciate Swiss chard, the alternative-food movement has long zeroed in on low-income, mostly black, urban neighborhoods like North Philadelphia, which they call food deserts because residents live more than a mile from a supermarket or large grocery store. As a result, people who live in those neighborhoods often rely on corner stores and fast-food restaurants, which don't sell much in the way of fresh produce or other whole foods. An army of advocacy groups has descended on these communities in recent decades, opening farmers markets, building community gardens, and endeavoring to bring a "farm to table" education into schools.

The impact of these efforts has been significant by many tallies.

These activists deserve at least partial credit for the overhaul of federal school-lunch nutrition standards that happened under Michelle Obama's watch in 2010. The number of farmers markets in the United States has nearly tripled, from just 2,863 in the year 2000 to more than 8,200 in 2014. There are more such markets in urban areas than ever before, and more than 7,000 of them feature vendors who accept federal assistance benefits (such as food stamps, officially known as Supplemental Nutrition Assistance Program or SNAP) as payment.

But the food activists' efforts have also had more complicated ramifications. Detoxing, clean eating, and a fixation on whole foods have replaced the calorie-counting, low-fat-yogurt dieting of the 1980s and 1990s. But they are just as much about restriction and rules as anything Jenny Craig ever told you to do. In fact, in 2009 Pollan followed his initial journalistic treatises with *Food Rules: An Eater's Manual*, which *The New York Times* described as "a useful and funny purse-sized manual that could easily replace all the diet books on your bookshelf." Note the specific use of "purse-sized"; with cutely judgmental advice such as "the whiter the bread, the sooner you'll be dead," *Food Rules* is the alternative-food movement's version of a women's fitness magazine.

And so, we buy organic blueberries at eye-watering prices to blend into our smoothies along with flax seeds, kale, and organic whey protein. We need to know how our meat was raised and where our chickens laid their eggs. We endlessly debate the benefits of full-fat milk, raw milk, and all manner of nut milks. I say "we," although it's often assumed that foodie obsession with sourcing and ingredients is the privileged preoccupation of the white, liberal elite. Yet in my interviews with half a dozen black women living in or close to poverty, I found that they were just as versed in this particular gospel, even if they weren't able to make the same purchases. Sherita talked about wanting to go to Whole Foods to buy "the fancy brown bread with nuts and seeds" instead of yet another stale white loaf from the dollar store. Tianna Gaines-Turner, a thirty-six-year-old mother of three and part-time security guard who has been homeless and who spent many years on some form of government assistance, told me she went vegetarian after watching the documentary *Forks Over Knives*, and was dying to learn to cook seitan. "But of course, I can't get that when I'm trying to make two hundred dollars in food stamps last through the month." The question is whether absorbing the alternative-food movement's brand of clean, whole eating has actually helped women like Sherita and Tianna—or introduced them to a new set of unattainable standards, driven by diet culture, now wearing organic farmer's overalls.

In asking this question, I do not mean to dismiss all that is hard and real about life in one of America's food deserts. Some 23.5 million Americans live in such neighborhoods, according to the USDA, and around 11 million of them are classified as low-income, because their household income is below 200 percent of the federal poverty level. Every time they want to put dinner on the table, these families face a tough choice: buy the limited selection of over-priced produce or cheap and readily available lower-nutrition fare offered in their neighborhood corner stores, or figure out the logistics of traveling some distance to buy more affordable "real food," as Pollan would call it, elsewhere. Finances make the decision even more fraught. If you're a single mother working three jobs, it's difficult to find time to travel out of your way to the better grocery store, let alone plan and cook meals. It becomes impossible if you can't afford to buy enough food in the first place or don't have a kitchen to cook it in.

Families like Sherita's and Tianna's are also disproportionately hurt by the health consequences of living with hunger. All the mothers I interviewed reported regularly skipping meals themselves in order to have something to feed their kids. But their kids were still vulnerable. Mothers in food-insecure households are three times more likely to report depressive symptoms as mothers who get enough to eat, according to Children's HealthWatch, a nonprofit that studies how economic conditions affect young children. Depressed moms are less likely to show affection, read stories, and offer other forms of interaction critical to a young child's developing brain; they're also more susceptible to job loss, drug addiction, and other crises.

When the situation is so dire that children do miss meals, the ripple effects become even more pronounced. "The scary thing is that a child's mental development will be impacted long before you see an effect on growth," said Deborah Frank, M.D., the founding principal investigator of Children's HealthWatch when I interviewed her for *Parents* magazine. When babies and toddlers don't eat enough, their bodies try to conserve heat for physical needs, so they become less active. "They sleep more and explore less, which means they aren't unpacking all the pots and pans in your kitchen cupboard, or experiencing other kinds of crucial learning that a normal child experiences," she explained. Older children who go hungry have higher risks of chronic conditions like anemia and asthma. They're also more likely to struggle in school or need to repeat a grade.

"You just sit there, thinking about food the whole time, and then you miss out in class," says Jahzaire Sutton, a fifteen-year-old high school sophomore who lives near Sherita in North Philadelphia. He and his younger brother have been raised by a single mother with disabilities; the family often struggles with food insecurity, especially toward the end of the month when their SNAP benefits run out. "I don't think people understand how much of a

toll hunger has on you mentally." He says in the past, some teachers came down hard on him for spacing out; he's not sure whether
they didn't understand what he was dealing with, or whether they
just felt helpless to do anything. One day in sixth grade, he did
tell his favorite teacher why he wasn't listening to her lesson on
figurative language in poetry, but it took courage. "I never want
to be a burden," he says. "But I have realized, kids need to speak
up and let people help us." The teacher gave him a granola bar. It
got him through the afternoon.

While it may seem paradoxical, hunger also plays a key role in
the development of obesity. "I often hear things like, 'Those people
can't be hungry—they're fat!'" says Janet Poppendieck, Ph.D., the
author of *Free for All: Fixing School Food in America*. "But the least
healthy, most obesity-inducing calories in our society are often the
cheapest." A study from the University of Washington found that
junk food can cost an average of $1.76 per 1,000 calories, while
more nutritious foods add up to $18.16 for the same amount. Food-
insecure families may also be more prone to obesity because their
bodies are essentially always in crash-diet mode, which ultimately
slows down metabolism. And all these effects on physical health
don't begin to get at how living in a food-insecure environment
changes your feelings about food. Forget teaching kids to honor
their hunger and fullness cues; these parents dread hearing their
child say, "I'm hungry!" So kids like Jahzaire learn quickly not to
say it.

They also learn to overeat when they have the opportunity,
which further distorts their ability to self-regulate around food.
Sherita tells me about a time when she worked in an office that
was staffed by a mix of upper-middle-class white women and low-
income black women like herself. "At lunch, all the Caucasian
people had tabbouleh or some vegan stuff. Maybe just a yogurt.
And all of us black women came in with these big cheesesteaks,"

she says. "The white women weren't eating for survival because they weren't hungry. So they could eat that healthy stuff. But we're hungry. We want big piles of food."

Those big piles of food may also offer protection, because, for many women, thinness carries an element of risk. In her book *Hunger: A Memoir of My Body*, Roxane Gay writes about wanting to make her body big to protect against violence and sexual assault: "I began eating to change my body. I was willful in this. Some boys had destroyed me, and I barely survived it. I knew I wouldn't be able to endure another such violation, and so I ate because I thought that if my body became repulsive, I could keep men away." Dieting might make you "white," as Sherita tells me her friends often joke about her obsession with salads and supplements. But it may also make you vulnerable in ways many women in these communities can't afford. The size-acceptance community often throws around the term "thin privilege" in arguing that thin people don't understand the oppression people with larger bodies face when shopping for clothes, flying on airplanes, or interviewing for jobs. But another facet of thin privilege is the level of privilege you must have in order to be thin safely. You must live in a safe neighborhood. You must be able to attract male attention without worrying that you'll be perceived as "asking for it." You must have the luxury of abundant food, if you are able to turn it down.

Shakirah Simley is a black food activist raised in working-class Harlem and educated in the Ivy League; she has spent the past decade trying to reconcile the aims of the alternative-food movement with the real needs she sees in low-income black communities— with the significance of that need for the cheesesteak that Sherita describes. One "aha" moment came early in her career, when she was working in public health in New York City, and spent an afternoon leading a cooking demonstration outside her church in Harlem. "I was making a watermelon, mint, and feta salad and people loved it," she says. "But then this dad came up to me and said, 'This is great, but my daughter says she's a vegetarian now and all we can get for her at Pathmark is potatoes.'" Feta cheese and fresh mint were not in the budget for a family who could not afford to splurge on luxury items that might go bad before they found time to cook with them. "I realized he was totally right," Shakirah says. "Adding one [farmers market] to this neighborhood is not going to cut it."

Shakirah followed that experience up by moving to San Francisco: "I thought, 'If I'm going to do this food stuff, I need to go where it's all really happening.'" She spent five years as the director of community outreach for Bi-Rite, a famed Bay Area chain of locavore-friendly grocery stores focused on supporting local farmers and artisanal food brands. "It's the grocery store of your dreams," she says; her job was to make sure the company knew how to connect with the local community when it opened up stores in working-class minority neighborhoods. Shakirah directed over $1 million of the company's resources into "anything that supported good, clean, fair food systems," as well as local community centers and youth programs. She also has her own artisanal food business, Slow Jams, where she makes preserves from free fruit scavenged around her neighborhood, as well as from produce sourced from local women farmers and farmers of color. When we speak, she's getting ready to come back east for a month-long fellowship at the Stone Barns Center for Food and Agriculture, the famous nonprofit that's also home to the celebrity chef Dan Barber's even more famous restaurant, Blue Hill, in Pocantico Hills, New York.

But Shakirah struggles with being a person of color in an overwhelmingly white industry. "I'm a black woman and I'm a food person, and seemingly, those two identities are separate," she says. Seeking advice from other black activists didn't initially prove useful: "I'd hear a lot of folks say, 'Okay, food is cute, but we're focusing on police violence right now.'" (Shakirah later updates me that the Movement for Black Lives has since added food to its official policy platform, in part thanks to an essay she wrote for the *Huffington Post* in 2017.) And her white colleagues didn't even seem aware of the problem. In their efforts to, say, bring farmers markets to disenfranchised neighborhoods, they would frequently talk about black people as hapless victims, cluelessly eating corner-store junk food and worn down by poverty and diabetes. That's because when white activists go into a black community, it is often with the assumption that they won't encounter anyone who is informed about healthy eating, cooking, growing food, or much of anything to do with the larger food system. "I am so tired of that narrative," Shakirah says. "Our food system only exists because of black people! American agriculture developed out of slavery. American cuisine comes from our foodways and our recipes."

Consider kale as Exhibit A: Today it's thought of as the whitest, most hipster of all vegetables, frequently served raw or in microgreen form. But it was originally grown by black farmers in the South, who cooked it in big pots with their collards, mustard greens, and ham hocks. "And yet we have all this white savior stuff, where people talk about going into poor communities to teach black and brown kids how to eat kale," Shakirah says. "That's the wrong approach." In fact, she argues, that approach is rooted in the same cultural oppression that underpinned many of the United States' early food programs. "Think of the federal lunch program; that was started by wealthy white ladies who wanted to take these poor kids, feed them all the same thing, and tell them this is how to be an American," she explains. "And going even further back, think about how feeding or not feeding slaves was a means of

controlling black folks. We have all of these dark historical prece-
dents for this kind of work, which nobody wants to look at."

I'm reminded of the first summer that Dan and I lived in the
Hudson Valley, when we joined our first community-supported
agriculture program. In CSAs, customers buy a "share" of a farm
for the season, paying up front so the farmers can afford to buy
seeds and equipment. The CSA members are then entitled to a
weekly haul of vegetables throughout the growing season. This
particular farm had done a fair amount of outreach to recruit low-
income members from Beacon, a diverse small city in Dutchess
County, New York, which has undergone rapid gentrification in
the past decade. At our first weekly pickup on the farm, a black
father stopped in front of a crate of bok choy and asked a volunteer what it was. "That's BOK CHOY," she responded, speaking
loudly and slowly, as if he might be deaf. She was white, in her
midforties, wearing Birkenstocks and a vintage sundress. "You can
chop it and sauté it. Or put it on the grill. Do you have a barbe-
cue grill where you live? Can you get some extra virgin olive oil?"

The black man's daughters ran cheerfully through rows of sun-
flowers with some other kids. There was nothing besides their
skin color to indicate that they might be members of a low-income
family; for all I know, he was a Vassar professor or an IBM executive. Whatever his socioeconomic status, there was no reason to
presume he had never heard of olive oil. But before we judge
that volunteer too harshly, I have to acknowledge that I sidled past
without joining the conversation—without offering that I, too, had
no idea how to cook bok choy.

I ask Shakirah why she thinks white foodies think they invented foods like kale and bok choy and assume that black folks don't
understand nutrition. She sighs. "Look, as much as I love Chez
Panisse, and the amazing work that chefs like Alice Waters have
done to bring this local, organic, French-inspired way of eating to

the American palate, there has been erasure there too." She points to Edna Lewis, who was born in 1916, the granddaughter of Virginia slaves, but grew up to work as a chef in some of the country's best restaurants and write *The Taste of Country Cooking*, a now classic cookbook on Southern cuisine that has become a best-seller again in recent years. Foodies love Lewis's descriptions of cooking fresh-caught trout over wood fires and measuring baking soda on coins during her rustic childhood. Lewis wrote knowledgeably about growing, harvesting, and foraging food, as well as cooking it. "Farm-to-table cuisine wasn't invented in Berkeley in the 1970s," Shakirah says. "But we have this cultural amnesia that allows all these white narratives around 'good food' to exist."

Erasure of black food culture has also come from another source arguably even more powerful than the white liberal elite: food marketers. Shiriki Kumanyika, a research professor at Drexel University's Dornsife School of Public Health and founder of the Council on Black Health, has been tracking the efforts of food marketers for over a decade. Her research shows that mainstream food brands target black and brown communities with messages they don't send into white markets. Among her most disturbing findings: African American and Hispanic kids see twice as many ads for candy and soda as their white peers. Media channels aimed at these audiences also show a higher percentage of food ads than those aimed at white or mixed audiences, and the foods advertised tend to be less healthy. Certain brands (one study pinpointed 7 Up, Kraft Mayonnaise, and Fuze Iced Tea) advertise only on Spanish-language channels, while others, such as Wendy's and McDonald's, devote significant portions of their advertising budgets to black or Hispanic media. Kumanyika says this isn't a conspiracy; it's just how marketing works. Part of the problem is surely that black media outlets sell more space to food advertisers because other kinds of advertisers—airlines, makers of computers and luxury cars—don't care about reaching a less affluent market. But either way, the result is clear: "Black people see all the ads targeted to the general population and then they also see ads targeted specifically to black people," says Kumanyika. "It's double the exposure."

Brands that sell well in black communities also invest heavily in those communities. McDonald's has sponsored "McTeacher Night" school fund-raisers in at least thirty states, while Pizza Hut rewards kids for reading with coupons for free pizza. Altogether, food and beverage companies spent $149 million on in-school marketing in 2009, according to a study in *JAMA Pediatrics*. Not all of that money was tied to educational enrichment, of course. So food activists fight hard against these programs; the Los Angeles Unified School District made headlines in April 2017 when it stopped allowing McTeacher Night fund-raisers. Parent groups across the country take pride in eradicating vending machines—but such efforts have been most successful in white districts. For neighborhoods that are otherwise severely under-resourced, a Kentucky Fried Chicken scholarship or Pepsi-branded basketball court is much less of a moral quandary. Kids need scholarships and basketball courts. Who cares who pays?

Underpinning all these efforts is the fact that marginalized populations tend to be more receptive toward advertising than white populations, notes Kumanyika. And the brands that invest this way are rewarded with strong consumer loyalty. This isn't because people of color have no interest in healthy food. In fact, Kumanyika is also quick to point out a long-standing tradition of community interest in what she describes as "alternative foods; not necessarily organic, but tied to health or healing in some way." It's because people of color are not used to being recognized for their purchasing power, or seeing themselves portrayed in positions of power, success, or health. "Ads [featuring black people] show us that people know we're important," says Kumanyika. "And the first brands to figure that out were soda and beer companies." To judge by Sherita's vision board—covered with white models because she can't find black women in fitness magazines—not much has changed.

Take all of this together, says Kumanyika, "and marketers are able to create a very strong following for foods that are laden with things that taste good and are affordable. All of which offers further reinforcement for a brand preference." And efforts to regulate these marketing decisions are quickly stymied by the political implications. "You can argue that we need to limit advertising to children because they are not yet at a stage developmentally where they can understand they're being marketed to and make those choices," Kumanyika explains. "But it's paternalistic to say to rational adults, 'We have to keep these products away because they aren't good for you.'"

That doesn't stop food activists from trying. In November 2016, the USDA released a report documenting how food stamp recipients spent their benefits; the media and advocacy groups alike seized on one small finding that people using food stamps spent around 5 percent of their food dollars on soda. *The New York Times* covered the story with the headline "In the Shopping Cart of a Food Stamp Household: Lots of Soda" and a photo of a shopping cart laden with Coca-Cola. Marion Nestle, a nutrition professor at New York University who is a prominent food activist, described the report as "pretty shocking," and called the government's Supplemental Nutrition Assistance Program "a multibillion-dollar taxpayer subsidy of the soda industry." This rhetoric fit nicely with the notion that poor Americans have been so ruthlessly brainwashed by food companies that they no longer know how to make good food choices.

In fact, a deeper reading of the report revealed that poor people don't buy that much more soda than other Americans, who spend 4 percent of their food dollars on it. We're all victims of insidious food marketing. Low-income black communities receive an even heftier dose, but this also comes with tangible benefits—representation and sponsorship—the value of which food industry critics haven't been able to fully appreciate, let alone replicate. The "white savior" approach to food activism is full of its own manipulative messages around good food, bad food, and good and bad people. And these conflicting food messages leave low-income families to continue the struggle.

"I have no idea why the government thinks the solution to hunger is a church basement food pantry," says Tianna Gaines-Turner. She and Sherita are both members of an advocacy group called Witnesses to Hunger, which was launched by researchers at Drexel University who figured that the best way to address issues of food insecurity, obesity, and malnutrition in low-income communities might be to ask the people who live there how these things work. Once you do that, you quickly realize why a lot of the solutions proposed by food activists and public health officials are met with blank stares. ("Don't get me started on community gardens," Sherita scoffs. "Black people eat meat!") A main point Tianna likes to make is that being hungry isn't really about hunger. If it were, maybe we could solve it through food banks, because it would just be a matter of getting more calories into people's homes. But that depends on their having homes with working heat, water, and electricity. It presumes that their hunger is a temporary problem, because they have jobs that will pay for the next round of groceries. And it ignores how much time and energy they're expending to fight much bigger battles, like Sherita's cocaine addiction, or

Tianna's daily fear that her black preteen sons will be gunned down in the street. Hunger becomes a kind of background hum against all of that. It makes everything harder. But it isn't always the first fire to put out.

There's also the grinding issue of stigma. Tianna was raised by a single mother who "did the best she could." But they often woke up in a house without heat, or were short on food. "And still, I got up every day and went to school and nobody knew," she says. "As a kid, I knew: You don't tell family business to outsiders, because at the end of the day, it is your business. You don't want someone to come and take you away because your heat is not on." That's a message she's passed on to her own three kids over the years, as she and her husband have struggled with job losses and medical crises that seem to pop up just about every time they're close to getting on track financially. "We teach them that what goes on in this house, stays in this house. You don't go to school and tell them you didn't have breakfast," she says, shaking her head. "But it's sad. We shouldn't have to teach our children that."

Like Sherita, Tianna lives in a run-down row house in Northeast Philly. She has a sign on the front door asking guests to remove their shoes; the carpet inside is spotless, though the family's belongings are piled up in every corner because there's not enough storage and no money to buy the organizational systems—baskets, bins, and labels—that upper-middle-class families consider an integral part of modern life. During one of my visits there is also a creeping patch of mold on the wall in the kitchen, which Tianna scrubs off every week because the landlord won't pay to fix it until the family catches up on the rent. But Tianna runs her household with scrupulous attention to detail. When her older son began to feel unsafe commuting to school, she pulled him out and found an online home-schooling program that could get him through middle school. "We'll try again when it's time to apply to high school," she says. She is similarly precise about monitoring her children's seizure conditions and arranging their after-school schedules. "I love you, make good choices, have a beautiful day," I hear her tell her daughter one morning when ten-year-old Marianna calls, per Tianna's rules, to say she has arrived at school safely.

On another of my visits, Tianna takes me with her to Walmart so I can see what it's like to shop on food stamps. In short: it's a ton of work. There is no browsing, no spontaneous splurging on ice cream, no trying a free sample and buying a fancy cheese this week just because. Tianna rifles through coupons and scrutinizes every single price tag: "This one's $2.50 but wait, wait . . . Here we go, a dollar on sale!" She loads up our cart with yogurts, bagged spinach, cereal, boxes of frozen waffles and vegetables, and fruit pops. The total comes to less than $100 for a cart full of heavy bags. Tianna is relieved; during her husband's last stretch of unemployment, the family of five received $793 in food stamps per month. But once he began making $12 an hour working nights at a meatpacking plant, their allowance was cut to just $200. Tianna makes $10.75 per hour at her job as a security guard, but neither of them gets full-time hours or benefits. In many ways, the math of their lives gets harder, not easier, with gainful employment. "We understand that we're both now making over the minimum wage," she says. "But it's just hard to have everything snatched away right when you're finally getting your foot in the door."

Back home, Tianna lays out her plans to cook in bulk over the weekend; she'll make pans of baked ziti and chicken that the family can eat all week. And she shows me the canned goods stockpiled in her kitchen, many of which were found at local food pantries in weeks when money was too tight for a Walmart trip. "Sometimes you just get garbage because people don't think when they donate, they just clean out their cupboards," she says as we unload the groceries. But Tianna knows how to get creative. Her favorite food-pantry score is a no-brand can of something labeled "beef with juice." She drains the juice to make gravy and mixes the beef chunks with mashed potatoes and string beans. "Yeah, I know it sounds gross, but it makes a meal. If we don't have something to eat, we can eat that."

I'm not sure what Michael Pollan would think of Tianna's shopping cart or her canned-beef recipe. There are certainly more packaged foods than whole, and there's nothing local or organic unless it was a random food-pantry find. Tianna herself is obese and has chronic acid reflux and asthma. And her own eating habits are sometimes chaotic. The day of our shopping trip, for example, she confessed to not having eaten; she said her stomach was killing her, but she was also making sure her kids got breakfast. Other days, she might make herself a bowl of lettuce sprinkled with orange juice for flavor. "I'm just not even hungry for anything else." But despite that type of necessity-driven restricting, Tianna has a strong grasp on nutrition, as well as plenty of ingenuity and an enviable knack for planning and cooking family meals. There is nothing passive or mindless about her food consumerism. She doesn't need to be lectured about the evils of fast food. All she really needs is to be treated with respect at the grocery store whether she's paying with cash or an EBT card, and to have enough money to feed herself and her family without worry.

"People eat better when they have higher incomes," says Julie Guthman, Ph.D., a geographer in the division of social sciences at the University of California–Santa Cruz. She has studied the race, class, and body politics of the alternative-food movement for over twenty years, and is perhaps most famous as Michael Pollan's harshest critic. Her 2007 piece "Can't Stomach It: How Michael Pollan et al. Made Me Want to Eat Cheetos" laid into the preachier aspects of the alternative-food movement and called out a certain hypocrisy when it came to understanding the true causes and implications of obesity. She also asked a pointed question that I've also often wondered about, as a not-exactly-thin person who eats a pretty "clean" whole-foods diet most of the time: "Why are Pollan, Goodall, and Nestle not fat? If junk food is so ubiquitous that it cannot be resisted, how is it that some people remain (or become) thin?" Guthman lives in Berkeley, adores farmers markets, and jokes that "I probably eat more like Michael Pollan than Michael Pollan does." The difference is that she's not convinced that their culinary preferences should be peddled as a universal solution. "We don't have to talk about how fat people are, or define what, exactly, 'healthy food' is according to some white-foodie standard. We just need to increase food assistance benefits and fair wages. That should be the message."

Guthman says Pollan and other leaders in the alternative-food movement have never publicly acknowledged her criticisms of their work. But in response to Donald Trump's presidential victory, Pollan, Mark Bittman, and others coauthored a new mission statement for what they call the "good food movement," which was published on the website Civil Eats in January 2017. They argued that "fighting for real food is part of the larger fight against inequality and racism," and they urged their movement to ally with others and fight for all progressive causes:

> You can't fix agriculture without addressing immigration and labor or without rethinking energy policies; you can't improve diets without reducing income inequality, which in turn requires unqualified equal rights for women and minorities; you can't encourage people to cook more at home without questioning gender roles or the double or triple shifts that poor parents often must accept to make ends meet. . . . The fight for healthy diets is part and parcel of these other struggles and it will be won or lost alongside them.

This represents a major shift from Pollan's 2008 open letter to President Obama, in which he argued for stricter regulations on food stamps so poor people couldn't use them to buy "nutritionally worthless foodlike substances" like soda, despite acknowledging that critics would say such a move "smacks of paternalism." But their new vision is short on specifics. How *are* poor parents supposed to work double or triple shifts and still cook family meals each night? And how do we factor in the emotional toll of these challenges, and the pivotal role food plays as a kind of solace? "When you've had a day and you've dealt with sexual harassment and microaggression and you're stressed out by finances and violence, you come home and you want to be comforted," says Shakirah. "You don't want to feel bad about what you're eating because there's already enough that makes you feel bad about yourself as a black woman. So who am I to say, 'Don't go to the Popeye's drive-through'?"

In her own conversations within these communities, Shakirah focuses less on what people eat and more on how they eat it; she emphasizes the importance of family meals, of putting away screens, of finding time to eat breakfast. "I want to sustain a culture around eating so people can recondition their relationship to food, but not in the kale smoothies–kombucha way," she says. I appreciate her desire to reframe the conversation, though I wonder whether rules obliging people to have family dinners and breakfasts aren't in some ways just as oppressive as rules about eating more greens. But Shakirah's larger point is that all these changes need to come from within a particular community; she can try to switch her mother to a lower-sugar brand of yogurt because she knows and loves her (and also knows that switching from Yoplait to kombucha would be several bridges too far). So perhaps it's a sign of progress that Pollan is being less prescriptive now. White activists can't target their advice as well or as sensitively because they don't know these communities' culture or history around food—and most of them haven't asked what people really want or need to be eating.

Jarrett Stein is trying to ask those questions—while fully acknowledging the limits of his abilities as a thin white food activist. He is one of the founders of Rebel Ventures, a nonprofit corporation supported by the University of Pennsylvania's Netter Center for Community Partnerships, where he works with his co-founders, two Philadelphia school system alumnae named Tiffany Nguyen and Corinthe Hamilton, and current high school students like Jahzaire Sutton to design, market, and distribute their own healthy packaged foods. In 2016, the students sold their first product to the Philadelphia public schools, a breakfast muffin called the Rebel Crumbles, which is now served in all the city's school breakfast programs. When I visit the commercial kitchen where Rebel Ventures rents workspace a few afternoons a week, I see half a dozen black teenagers wearing aprons and hairnets, industriously chopping bananas and measuring out oats and apple juice. Jarrett wears the same apron and hairnet, and somehow looks right at home, despite the incongruity of his upper-middle-class upbringing in Nashville, Tennessee, and the private-school education that culminated in an Ivy League degree. "With Jarrett, everything has to be very healthy," a student named Fleshae tells me, with an affectionate eye roll.

But Jarrett comes by his passion for healthy food honestly; at age fourteen, he started losing weight and having trouble breathing. His mother got worried when playing a game of tennis left him coughing and wiped out. Jarrett's pediatrician detected a heart murmur and referred him for an echocardiogram, which the radiology team couldn't complete because they found a tumor in his chest so big that it had pushed his heart to the right side of his body. Jarrett was diagnosed with non-Hodgkins lymphoma and immediately admitted to the hospital for three weeks of treatment. Then he began a six-month course of chemotherapy.

Chemotherapy sapped Jarrett's appetite and he began losing weight rapidly. "The doctor was like, 'Look, we're going to have to put you on a feeding tube unless you can get over the chemo nausea and start eating,'" he recalls. "I knew on a tube, I wouldn't be able to eat like a normal person. The thought of losing control of what, when, and how I ate was devastating." Still, Jarrett felt terrible. He was stuck at home, missing school and his friends, unable to do much more than watch television for hours every day. So he started watching the Food Network. And then he decided to start cooking. "My parents spoiled me a ton; we'd go to Williams-Sonoma and buy a pasta machine so I could make fresh pasta for dinner," he says. "So then I had all these ridiculous gadgets plus all of this time, and I started to really like cooking." It wasn't just about eating "healthy," though he began to get pretty excited about that too. Food brought Jarrett back to himself. Jarrett's bio on the Rebel Ventures website begins: "I am from sweet tea, smiles, truth and hurt. I am from the grass beneath bare feet. I am from the hospital. Sterile walls, sterile smells, and the drip, drip, drip of chemo. I am from hope. . . . I am from a deliciousness called healthy."

By the time Jarrett was fifteen, his cancer was in remission and he was back at school, writing a food column for his high school's newspaper. In 2005, as a freshman at the University of Pennsylvania, he took a class on the psychology of food taught by Paul Rozin, Ph.D., one of the world's foremost researchers on food. Rozin famously coined the phrase "the omnivore's dilemma" that inspired Pollan to write his best-selling book. "Paul was as in love with food as I was, so I joined his research lab, but we really just talked about

food we liked and went out to eat a lot," Jarrett says. He went on
to take every class Penn offered with "food" in the name, including "The Politics of Food" (taught, incidentally, by my stepmother,
Mary Summers, who is a senior fellow with the university's Fox
Leadership Program and a lecturer in political science who teaches
academically based service classes). That class required Jarrett to
volunteer in the Philadelphia school system, and brought him face-
to-face with the dire reality of school food and nutrition in low-
income neighborhoods, where Williams-Sonoma pasta machines
might as well be alien spaceships. For someone who felt such a
personal connection with food as a means of comfort and healing,
it was a crushing realization. "Let's just say it was not what Michelle
Obama wanted," he says. "The kids are bringing in all of this stuff
from the corner store because that's where their real education
about food is happening. And if anything, the way we're teaching
nutrition in schools is pushing them towards that."

In his first year out of college, Jarrett got a job teaching nutrition classes to Philadelphia middle school students. He was assigned
to four different schools and visited each once a week to teach what
he describes as a "government-approved nutrition lesson." He was,
by his own admission, terrible at it. "I didn't have relationships with
any of the students, and the curriculum was very top-down and
didactic about what kids are supposed to eat," Jarrett explains.
Without quite realizing it, he had signed up for one of the
alternative-food movement's suicide missions, preaching about
food groups and calories to kids who didn't ask to be saved.

On his first day at Vare Middle School, Jarrett was sent to the
physical education teacher's classroom. It had bare walls and few
supplies. The teacher—a white man in his thirties—explained that
he hardly ever used it because "kids don't want to learn about
health." He preferred to let them play basketball in the gym all
period instead. Jarrett had brought a bag of apples along for his

presentation and, trying to forge some kind of bond, asked the guy if he'd like one. "He declined and told me he was on a diet he learned about from an infomercial, where he could only eat two special cookies a day." Then their first class arrived and Jarrett realized that his colleague's lack of nutritional awareness would be the least of his problems. There were thirty eighth-graders in the class, and he and the gym teacher were among the only white people in the room.

Jarrett was nervous, but he had a lesson plan to follow. He went to the front of the classroom with his bag of apples and greeted the class: "Hello everyone! Today we will learn about apples." Then he began reading a history of the fruit. He was immediately interrupted: "Yo, you didn't even tell us your name." Jarrett apologized and regrouped. But a minute later, another student's hand shot up. "Mr. Stein, don't you know we don't give a fuck about your white-people food?"

The class erupted in laughter and then chaos. Kids were out of their seats, running into each other, yelling, and throwing Jarrett's apples around the room. The PE teacher took over: "To the gym!" The class cleared out. One girl lingered behind to say, "These kids are so disrespectful. Can I have an apple?" Jarrett gave her one. Then he picked all the apple chunks up off the floor and wondered what the hell had just happened.

All children are, of course, influenced by their exposure to food in school. But for low-income kids, the relationship is especially profound because their free or low-cost school breakfasts and lunches may constitute most of their eating for the week. If there's not much food at home, all of your ideas about what a healthy meal looks like will come from that cafeteria tray. Janet Poppendieck suspects this has played a role in attitudes about what Sherita calls "white-people food," because when schools preach too much about what kids should eat, it tends to harden their lines of defense.

"It gets your back up, to have somebody tell you that something you want to eat isn't good for you," Poppendieck notes. And delineating good and bad foods has been a big part of school food's complicated history.

From the early 1940s until the mid-1970s, the federal standard for school nutrition was something called the Type A meal. The moniker comes from the National School Lunch Act of 1946, which guaranteed federal cash subsidies for school lunch programs. In order to ensure that the government funds were put to good use, the legislation included guidelines about what a school lunch had to look like in order to qualify. The Type A meal specified exact serving sizes of five components: a meat, a grain, some kind of dairy, a fruit or vegetable, and, initially, two pats of butter. (Schools also received smaller amounts of funding for Type B meals, which were more like snacks, and Type C, which was just milk.) "The Type A meal was what good nutrition looked like in 1946, and you have whole generations of school lunch directors who developed their menus based on that model," explains Poppendieck. Sometime in the 1970s, critiques began and the Type A meal was tweaked. "Meat," for example, was expanded to include high-protein alternatives like peanut butter and eggs. But school lunch directors remained beholden to that template because they feared they might lose funding if their meals were audited and found lacking in one category or another.

Poppendieck credits the rise of processed foods in school lunches to both the Type A meal and the USDA's Child Nutrition Labeling Program, which lets food manufacturers designate certain products as meeting the school lunch nutrition standards—allowing them to market to school lunch programs—as long as they're willing to bear the financial responsibility if government auditors find their goods lacking in any way. "People running cafeterias have always been very low-status in the school system and they had a lot of fear and anxiety about meeting those standards," she notes. "So when industry came in with things like Hot Pockets, which have labels promising that they meet government standards, it made life much easier for food service directors." But as school lunches became increasingly processed, they also became a lightning rod for the alternative-food movement and other critics. In the 1990s, school menu planning shifted from Type A meals, which Poppendieck calls a "food-based model," to a "nutrient-based model." Under the new system schools instead had to ensure that each meal contained specific quantities of protein, carbohydrates, fat, and other nutrients. "In wealthier school districts, where you'll find a school dietitian who knows how to use computer modeling software, this may well have enabled an escape from the tyranny of processed foods," Poppendieck says. "But it also drove many systems toward using even more prepackaged items and toward contracting their meal services out to large companies." Aramark and other corporate food suppliers were already set up to calculate protein grams and other nutrient information, while a small-town cafeteria director who began her career serving in the lunch line might never have used a computer.

Another nutritional mission gone awry is the story of fat in school lunches. The 1990 Dietary Guidelines (a set of nutritional advice issued every five years by the federal departments of Health and Human Services and Agriculture that serve as the foundation for all federal food and nutrition programs) advised that no more than 30 percent of a child's daily calories should come from fat. "This meant school nutritionists were suddenly trying to produce meals with only 30 percent of calories from fat, which is really difficult to do if you also have to maintain a calorie floor," says Poppendieck. "After all, whenever most of us try to restrict our fat intake, we do it by cutting calories." Since that wasn't an option for school meals, food service directors and dietitians compensated by replacing calories from fat with calories from sugar. Desserts were added to menus. Plain whole milk was replaced by chocolate or strawberry skim milk. In 2005, the government's Dietary Guidelines increased the goal from 30 percent to 35 percent of total calories from fat. And after Michelle Obama's school lunch overhaul of 2010 (which, Poppendieck says, brought menus back toward the food-based approach of yore, albeit with some modern nutrient-specific requirements), the 2015 guidelines eased up about fat considerably, specifying only that saturated fat should be limited to 10 percent or less of daily calories. But the long reach of those 1990 guidelines can still be felt in many cafeterias, and in the palates of a generation of Americans who grew up preferring intensely sweet flavors.

The night after his history-of-apples lesson, Jarrett couldn't sleep. "I hated that I just froze," he says. He was used to thinking of himself as a resilient person—he was the kid who beat cancer! But he was suddenly unsure of his purpose, and didn't know how he could reframe his failure in that classroom. He wanted to share his passion for food with these kids, but how was he supposed to connect with them or teach them anything useful about nutrition? The fix was not instant. Jarrett wasn't able to connect with all the kids he taught that year; mostly, he says, he continued to learn by failing, week after week. But he also began to figure out that what he really had to do was engage the kids as experts on their own eating. The dogmatic, Dietary Guidelines–driven approach was never going to work—especially when those same guidelines helped account for why these kids were turned off by healthful food and drawn to the super-sweet offerings of their corner stores. He arranged the desks in a circle, instead of in rows facing him. He threw out his lesson plans and asked the kids to talk about what they would stock if they ran their own corner store or supermarket. He paired them up to debate questions such as "Is your school healthy?" And he asked them: How would you create a healthier school? What would you want to eat? One group of middle-schoolers was so inspired they decided to start making their own food. When those kids went to high school, Rebel Ventures followed.

Eight years later, it's become an independent nonprofit food business. Jarrett sets out a To Do list at the start of each work session and does a little low-key cruise directing to get people on task. But he wants the kids to take total ownership of the business, the recipes, the food. The Rebels donate Crumbles to fruit stands hosted by schools around the city. They also conduct peer-led "tasting team" workshops in classrooms and cafeterias, where kids get to try new foods and rate them. The Rebels are paid a competitive hourly wage for their efforts, so the job enables Jahzaire to help his mother pay the rent. And they're learning the basics of entrepreneurship, marketing, sales, and design, along with nutrition and cooking skills.

In order to develop their Rebel Crumbles for the school breakfast menu, the group studied the school district's nutrition criteria and decided what recipes to develop and what kinds of ingredients to try. For school food directors, the Rebel Crumbles offers the same perks as a Hot Pocket: It's a complete meal, meeting all the school breakfast nutrition requirements, in one convenient, ready-to-eat package. But the ingredients are certainly higher quality: one Crumbles packs in a half-cup of apples and cranberries along with 16 grams of whole grains. It also contains far more sugar than Marion Nestle or Michael Pollan would want to see in a breakfast food. But Jarrett is clear about why this sugar is there. "This was

created by kids, for kids," he says. The Rebels taste-tested fifty versions of the Crumbles before they hit on one they wanted to eat.
It was only after they began to translate those recipe tests to a nutrition label that they realized how much sugar was involved. Experiments with alternative recipes are now under way; on the day I
visited, the kids were using apple juice as a substitute for sugar.
Jarrett insists that any such changes need to come from the kids.
"The reality right now in schools is that lots of kids are taking
cereal and apples for breakfast, and throwing their apples away,"
he says. "We need this to be something kids will really like." Fleshae agrees. She was part of the team that took early Rebel Crum-
ble samples into elementary schools for taste tests. The younger
kids repeatedly told her to make the muffins sweeter. "If kids find
out this is healthy, they won't want to eat it," she tells me, like the
clever food marketer she has become.

Rebel Ventures is easy to love. After all, it's motivating kids to
learn about health! And cooking! And business! It offers the kind
of story adored by local news channels and charitable foundations
alike. And Jarrett hopes it will also be easy—or at least feasible—
to someday scale up to other school districts, in other cities and
other states. But if that is to happen, food advocates, as well as
school administrators and school-food policy makers, have to be
ready to embrace his community-driven model. They'll have to
accept more sugar than is nutritionally optimal as a reasonable
trade-off for letting teenagers be the true experts on how to feed
themselves. For a movement that has organized itself around a
series of gurus and their philosophies and rules, that part might
not be so simple. It speaks to so many bigger cultural needs, like
the need to separate messages about thinness as a beauty ideal
from conversations about diabetes and blood pressure, and to
accept that people who don't look like our picture of health can

still be authorities on their own bodies. The need to stop viewing processed foods as not-foods and start understanding the significant roles they play in people's lives. And the need to end the white savior model of food activism and replace it with something more authentic. If we could do all that, Sherita might still have skinny fitness models taped to her vision board. But maybe they could look a little more like her.

Six

Bypassing Hunger

I'm in a dietitian's office, located within a bariatric surgical suite, housed on a quiet floor of a large hospital in Boston, Massachusetts. The room is tiny, with bad fluorescent lighting. It's dominated by a large bookcase filled not with books but with empty yogurt containers, flattened granola bar boxes, and rinsed-clean protein-shake bottles. It's as if somebody went on a very low-calorie snacking spree and then preserved all the packaging as a souvenir. The dietitian who works in this office uses these packages to illustrate to clients how to put together what she considers a proper meal. But this particular client, a thirty-six-year-old special-education teacher named Gina Balzano, doesn't really need the lessons.

Gina, one year out from her gastric bypass, has invited me along on this checkup so I can see how it feels to be an "After" in the world of weight-loss surgery. Her dietitian, whom I'll call

Ramona, is a tall, thin woman dressed in shades of brown. "She's the one we're all scared of," Gina tells me before the appointment. "We" are the other patients she bonded with in her support group. And I can see why Ramona triggers their anxiety; she's unsmiling and pecks skeptically at her keyboard as Gina lists out everything she now eats in a day. Breakfast is an egg muffin, a batch of which she makes once a week from half a dozen eggs and assorted vegetables. Around ten, she eats one Chobani Simply 100 Greek yogurt. "Do you know that brand?" Gina asks Ramona, who nods, flicking her eyes over to the case of yogurt cups. "Oh right, of course, you probably have five over there."

Lunch and dinner are Blue Apron meals; Gina cooks one each evening to share with her husband, then packs up the leftovers for the next day's lunch. "I get meals that are between five hundred and a thousand calories for a traditional portion and then I third it for myself," she says. "I eat three or four ounces of protein and the vegetables, and maybe a tablespoon of the starch. I don't like to waste room on the starch, I'd rather have the vegetables." Her afternoon snack is half an Atkins protein bar.

Gina also admits to having had two alcoholic drinks in the past month, which Ramona records with a frisson of disapproval. Gina tells me later that she also eats a small square of dark chocolate every night. "Because I want to feel like a human," she says. "But I didn't think Ramona needed to know that." Gina doesn't track her total calories, but in general, she shoots for half the portion size listed on any food label. Before her surgery, Gina usually ate three big meals a day: a breakfast sandwich and coffee from Dunkin' Donuts, some other kind of fast food for lunch, and then a home-cooked dinner. "Now I eat smaller portions but way more regularly," she says. "It is so much better than being hungry all the time or gross full right after I eat, and then starving for hours." But although she eats more frequently, her total caloric intake is almost certainly lower: Most weight-loss-surgery patients at Gina's stage of the process are told to eat between a thousand and fifteen hundred calories per day.

"That's great," says Ramona, typing away. "Now you might think about playing with your dinner proportions. Like, you might do just two ounces of protein, instead of three to four, and make up the rest with more vegetables."

"Okay, great," says Gina. "I love vegetables. They help my stomach."

"Well, and the thing is, I otherwise see people's proteins starting to creep up," says Ramona. "And it might get to the point where—"

"Where I feel uncomfortable?" Gina asks. Also on the "Don't Tell Ramona" list is the muffuletta sandwich she ate a few months ago, on a weekend trip to New Orleans. Gina ate more than she normally would because it was so delicious; an hour later, her newly shrunken stomach sent it all back up.

"Well, maybe uncomfortable. Or you might not even feel uncomfortable, that's the point," says Ramona. "People start being able to eat a little more now. And I'm sure they told you that everybody hits a low point, a nadir with weight loss, so some regain is normal. That's why you might want to overshoot a little now. I mean, not really—but because that way you can creep back up and still be at your goal weight." Ramona talks quickly, in a kind of medical shorthand that Gina has become accustomed to over the past year of these meetings. So it takes me a minute to realize that this is really advice on how to rig the game in Gina's favor: "Lose more weight than you need," Ramona is saying, "so when you inevitably regain some, you don't end up right back where you started." It's like a dieting insurance policy, or the kind of tip I've seen (and probably written) in many a women's magazine article. Ramona types some more. "Like, it's good that you just eat half the protein bar for a snack. I think that's smart."

"Well, the whole thing feels like too much," offers Gina.

"It is too much!" Ramona says. She is suddenly fierce. "I see people all the time, eating the whole thing just because they think it's a good food. Just because you can or it's good, doesn't mean you should."

She scrutinizes Gina's food log some more. "So you're busy. You work a lot. Right? That's why you use those Blue Apron meals?"

Gina acknowledges that she's busy, but adds, "I actually think it helps to be busy because I have to plan ahead. I bring my breakfast, lunch, and snacks to work and I'm good to go. And the Blue Apron meals make it really easy to comply."

"No, I just ask because you may get to the point where you can do it all in your sleep, without that," says Ramona. By "it all" she means meal planning, grocery shopping, and cooking from scratch. For all the packaged foods littering her office, it seems that Ramona wants her clients to be cooking and eating whole foods— never mind that, as Gina gently points out, Blue Apron meals do require cooking, if not planning and shopping. I wonder if that's because a diet charted out by a delivery service and interspersed with processed convenience foods falls short of our current cultural ideals about "clean" eating. Ramona doesn't want her clients merely piecing together their diets from her cabinet of packages as if assembling a jigsaw puzzle. She wants them confidently navigating the entire landscape of food, living life to the fullest—while still only consuming two ounces of protein at a time. She tells Gina: "I'll give you a cookbook that we made here; it has some great recipes."

After the food discussion, it's time for Ramona to present Gina with her Before and After photos. The Before, taken on August 26,

2015, shows Gina, who is five feet, four inches tall, weighing 340 pounds. She wears a loose white tank top and long, baggy black shorts and stands unsmiling against the office wall, resigned to the humiliation of the photo. In today's After, she weighs 187 pounds and wears a short purple paisley shift dress with black tights and boots. She's smiling in this shot, still self-conscious about the ritual, but more amused by it this time. And even though, technically, Gina is still obese—her body mass index is 32.1, over the "obese" cutoff of 30, though well down from her original 58.4— the difference in the photos is pretty staggering. She's lost twenty inches from her hips, nineteen inches from her waist, and three inches from her neck. Earlier in the appointment, a nurse calculated that this means Gina has lost 82 percent of her "excess weight." That's a term only loosely defined in the weight-loss-surgery literature, because deciding how much of anyone's body mass is "excessive" is fairly arbitrary. To put it more accurately, Gina has lost 45 percent of her starting weight. But whichever stat you use, across the office, she is considered the current superstar. It's a role she's not entirely comfortable with. "I actually hate Before and After pictures," she tells me in an email a few weeks before the Ramona meeting. "It was work to love myself before surgery and it's work now. I am a During, at all times."

But Ramona loves the photos. "Wow, look at that!" she says, and her face lights up, becoming more animated than she has been for our entire meeting. "Oh, that's what gets us up in the morning!" She brings the photos over to the nurses' station and passes them around. Everyone oohs and ahs at the newly trim Gina. "Isn't that fantastic? That's amazing. That's beautiful. Congratulations, Gina."

We say goodbye and head to the elevators, where Gina's husband, Nick, is waiting for us to go to lunch. "I do look really different and in a lot of ways, I am different," Gina says. "But that felt like she was so happy because now she can look at me and not feel sick."

Bariatric surgery has long been considered the shameful and risky last resort of the "severely obese," a category that the Centers for Disease Control and Prevention defines as including anyone with a BMI of 40 or more. The other term that medical professionals use to describe these patients is "morbidly obese." By invoking morbidity, doctors frame bariatric surgery as a critical health-saving strategy—necessary when someone is too fat to walk, work, or live a normal life—even though from 1995 to 2004 almost 1 percent of bariatric-surgery patients died within the first year post-op, and nearly 6 percent within five years, according to data published in the journal *JAMA Surgery*. It was a mortality rate verging on what medicine traditionally considers an unacceptable level of risk for any surgical procedure. But a recent refinement of surgical techniques and medical care have brought the one-year mortality risk down to 0.11 to 0.23 percent, depending on the procedure, according to data from 2008 to 2012 published in the *Journal of the American College of Surgeons*. And the procedure's popularity is rising correspondingly; surgeons performed almost forty thousand more operations in 2015 than in 2011.

Altogether, 196,000 Americans had some form of weight-loss surgery in 2015. More than half of them had the same procedure as Gina: a sleeve gastrectomy (often referred to as a gastric sleeve), in which a surgeon removes 70 percent to 85 percent of a patient's stomach, then staples what's left into a small banana-shaped pouch. Another 23 percent underwent the Roux-en-Y gastric bypass procedure, in which a walnut-sized section is removed from the original stomach and connected directly to the intestines. Food lands in the new, smaller stomach, while the original organ floats nearby, unused. (Less than 6 percent of patients had the once popular adjustable gastric band surgery, in which the stomach remains intact but is corseted by a removable, inflatable band; the remaining cases mostly involved removing gastric bands or converting them to the permanent gastric sleeve.)

In addition to becoming less deadly, proponents of bariatric surgery say it seems to be working better than it did in the past. While the original gastric band surgery was associated with slow weight loss and frequent weight regain (especially once the band was loosened or removed), patients who undergo a Roux-en-Y or gastric sleeve procedure generally lose 25 percent to 35 percent of their original weight. But it's important to frame that success rate carefully. Losing a quarter or more of your body weight may sound huge, but "most people stay well within the range of what we consider obesity," notes Arya Sharma, M.D., Ph.D., an obesity researcher at the University of Alberta and the scientific director of the Canadian Obesity Network, a consortium of more than fifteen thousand obesity researchers and health professionals. It's also unclear how long the weight loss is sustained; as Ramona warns Gina, the data also shows that some weight regain is inevitable. And no one knows much about the long-term impact of these surgeries, because no randomized control trial has been able to follow an entire cohort for more than a few years. One of the biggest, the Swedish Surgical Outcome Study, followed 2,010 surgery patients for almost two decades—but the participation rate dropped to 84 percent after the first ten years, and to 66 percent by the fifteen-year mark. This hampered the researchers' ability to draw firm conclusions, because nobody knows what happened to the other 34 percent.

But even with all these qualifications, the evidence is clear that surgery offers a more significant and perhaps more durable weight loss than any drug or diet on the market. As we'll see in this chapter, it does so through a series of biological mechanisms that can totally transform how a person relates to food, in surprising and nuanced ways. Bariatric surgery seems to rewrite our eating instincts, changing a patient's experience of hunger and fullness so much that Gina really can be done after just half a protein bar. To Ramona, to Sharma, to anyone else working on the front lines of the "war on obesity," and to many patients who have undergone the procedure and lost weight after decades of failed diets, this is a revelation. But others challenge the premise that fat bodies need to be cut apart and redesigned to make them smaller. And there is something fundamentally flawed about the notion that people need to be freed from their original appetites and taste preferences, especially because sustaining this so-called freedom requires patients to follow all sorts of rules and rituals about what, when, and how often they can eat. "Bariatric surgeons are prescribing for fat people what we diagnose as eating disordered in thin people," says Deb Burgard, Ph.D., a psychologist in Cupertino, California, who specializes in eating-disorder treatment. She's also a longtime activist and a co-founder of the Health at Every Size movement, which argues that our culture's fixation on weight loss is discriminatory and entirely the wrong way to go about improving public health. "They don't understand the trauma faced by people living in higher-weight bodies and they are not thinking about their role in that trauma. They just see a fat body as proof of an out-of-control hunger, and believe that getting rid of that hunger is the solution. But why would never experiencing hunger be a good thing?"

When Burgard poses that question, I am initially stumped. After all, I lived for two years with a child who never experienced hunger. It was not a good thing. It broke my heart daily. And without the safety net of a feeding tube, Violet's lack of hunger would have killed her. We are born with the ability to experience hunger because eating ensures survival. But I also know—because I live in the world—that resenting and regretting our hunger has become part of the normal business of living, especially as a woman, especially right now. We apologize for taking a cupcake at the office party, whether we're truly remorseful or just feel expected to perform our penance. We skip breakfast, yet feel annoyed when our stomach is rumbling for lunch at ten a.m. We go to dinner with friends and order the salad or don't order the fries because we're trying to match our appetites up with what everyone else seems to be doing. We joke that we'd never want to be anorexic, but gosh, we admire that willpower. And anyone who lives at a higher weight knows that how he or she displays and responds to hunger will open them to judgment, curiosity, ridicule, stigma. *Why would never experiencing hunger be a good thing?* For many people, it would be living the dream.

We don't want to be hungry because our culture has told us that we don't want to be fat. Sixty percent of Americans are currently trying to lose weight and 75 percent have made some effort in the past, according to a survey published by University of Chicago researchers in October 2016. And there is a deeply held belief in our society—one that runs all the way back to the Bible, to the seven deadly sins—that people get fat because they are gluttonous, slothful, and weak, and lack willpower around food. This isn't true: Though some obese people do eat compulsively (as do some thin people), the vast majority do not. Only 3.5 percent of women and 2 percent of men are diagnosed with binge eating disorder (itself a complicated psychological condition that is about much more than self-control), while 68.8 percent of Americans are classified as overweight or obese. Even if binge eating disorder is wildly underdiagnosed, it's a crude mischaracterization to assume that being overweight is only about eating too much. Genetics, biology, psychology, socioeconomic status, and other environmental factors all contribute to body size. "We know there are probably a hundred or more kinds of obesity, each with different causes and clinical characteristics," says Lee M. Kaplan, a gastroenterologist and the director of the Obesity, Metabolism, and Nutrition Institute at Massachusetts General Hospital. Burgard argues that even attempting to classify obesity by type or origin is misguided: "We have this fundamental misunderstanding that everyone should be close to the same weight, and therefore higher weight bodies can never be healthy and well regulated," she explains. "But what if most people's bodies are regulating themselves fine, just at a wider variety of weights than we've been taught to consider acceptable?"

Nevertheless, the willpower misconception persists, and it contributes to our sense that being overweight is dysfunctional and abnormal—that the size of our body is proof that our eating is somehow out of control, and that we'll only have a good life if we can conquer our hunger and lose the weight. Because we think hunger is bad and weight loss is good, the idea that a surgery can remove the former and achieve the latter is deeply seductive. But one consequence of that trade-off is never again eating the other half of the protein bar, let alone the muffuletta sandwich. Is merely removing the experience of physical hunger enough to cancel out that loss? Can someone's ability to eat really be so permanently transformed? The very reasons for weight-loss surgery's purported success also require us to ask: Should we be doing it at all?

Gina and I grew up a year apart in the same midsized wealthy Connecticut town. We were friendly in high school because we were both bookish theater nerds. We lost touch in college, but more recently have run into each other once a year because we frequent the same annual knitting festival.

Although we were raised in the same place, we had different childhoods, because Gina comes from one of our hometown's few working-class families. Her father was a gunsmith with an eighth-grade education, and her mother drove a school bus. "We were very, very poor," she tells me as we all order lunch at a Boston pub after the meeting with Ramona. I get a burger with fries, and Nick orders fried chicken. Gina, who went online and studied the menu ahead of time to figure out what she can eat, chooses two appetizers. "And when you're poor *and* fat, you're just so screwed." Gina was chubby as a little kid and obese as a teenager; everyone in her family is overweight to some degree. Gina used to have to untie her father's shoes for him at night because he was too tired to bend over and reach his feet. She didn't realize that was weird until she told a skinny friend, who said she never had to help her own dad like that.

Gina says that on family grocery store trips she never begged for cookies or chips; she wanted oddball things like salsa or canned asparagus. "In retrospect, I think I really craved fresh fruits and vegetables, but I didn't know how to articulate it. Everything my mom got was canned because we couldn't afford to buy fresh. She was feeding us on like, forty dollars a week." The family ate lots of venison because her dad was a deer hunter. On special occasions, there would be her grandmother's homemade lasagna. "She'd be cutting me a giant portion and telling me to eat it in the same conversation where she told me I was too fat," Gina recalls. And then there were late-night binges with her dad, during which they ate entire sleeves of Oreo cookies. "I got a lot of mixed messages about food."

Gina had a childhood friend who lived in the neighborhood and was also poor and overweight. Gina would go to her house after school and eat hot dogs stuffed with cheese. "I would have three," Gina remembers. "So there was a reason why we were fat. We were not eating healthy, ever." Gina and her friend went on their first diet at age nine, following a plan from the friend's doctor—just a one-page list of foods they could eat. Gina stuck with it as much as she could for three or four months. "But there wasn't a lot of follow-through for diets at my house," she says. "Like, my dad probably made a ridiculous amount of cream puffs or fried dough pizza, and we just kept eating like that." Next she joined Weight Watchers—with her father this time, although their progress was once again quickly derailed. "After that, I was pretty much on and off diets all the way through college," Gina says. She lost some weight on each diet, but never got thin. The least she ever weighed was 198 pounds, when she was twenty-one and had just moved back to her hometown after college. She was miserable and unsure about what she was doing next, so she started the South Beach Diet and began going to Jazzercise classes three times a day. "I think I stayed at that weight about three days," Gina says. "Then my friend's mom died and I ate chips again."

Gina stopped dieting a few years later, after a stint with "one of those strip-mall diet places" where she went for weekly weigh-ins during which a "nutrition counselor" would go over Gina's diet and then sell her on their protein powders and detox juices. Initially, as always, the diet worked. In fact, this one really worked: Gina lost 80 pounds, getting down to 212 pounds, one of her lowest weights ever. But she was still in the "twenty-something flail period" as she calls it, meaning she was single, broke, and trying to figure out how to get her career off the ground. "I was finally losing a lot of weight, but I was still miserable," Gina says. "And it was like, 'Holy shit, being thin doesn't make you feel better?'"

Gina started to research the diet industry and learned that brands like LA Weight Loss and Weight Watchers build their customers' diet failures right into their business models. "Every single time I lost weight on a diet, I gained more back," Gina says. "Every single time." She's not alone: Weight Watchers' own research, as reported in *The New York Times Magazine*, finds that the average customer will lose 5 percent of her body weight in six months, but regain a third of that weight within two years. For someone starting at 200 pounds, that means a net loss of less than 7 pounds. Only 16 percent of Weight Watchers customers maintain their weight loss for five years, according to a 2007 study published in the *British Journal of Nutrition*. "[The company] is successful because the other 84 percent have to come back and do it again," a former Weight Watchers executive named Richard Samber told interviewers in the 2013 BBC documentary series *The Men Who Made Us Thin*. "That's where your business comes from." Weight-loss success rates are similarly dismal for other popular diets like South Beach, Atkins, and the Zone; while most dieters lost around 10 pounds in the first six months—a hollow victory, right there, for many people—virtually everyone regained some of that weight within two years, according to a 2014 review published in *Circulation*, an American Heart Association journal.

The data isn't any better for the newer, trendier, less diet-y diets. David L. Katz, M.D., is the director of the Prevention Research Center at the Yale University School of Public Health and himself the creator of several popular programs, including *The Flavor Point Diet*. He's the kind of high-profile, well-credentialed obesity expert that women's magazines and morning shows love because he's charming and knows how to translate scientific literature into digestible sound bites. And he gives the kind of optimistic advice that makes weight loss sound infinitely doable, if only we would remember to start our days with bowls of Greek yogurt and berries and end them with kale salads. (Katz and his wife also launched a healthy-cooking website called Cuisinicity, the tagline of which is "Love the foods that love you back.") So it's fair to say that Katz is pretty pro-diet, or at least, strongly in favor of rules around what we should and shouldn't eat. And yet, when he compared research on low-carb, low-fat, Mediterranean, Paleolithic, and vegan diets for a 2014 paper published in the *Annual Review of Public Health*, even he was forced to conclude: "Can we say what diet is best for health? If diet denotes a very specific set of rigid principles, then even this necessarily limited representation of a vast literature is more than sufficient to answer no."

Diets don't work because they require us to live in a constant state of war with our bodies. "Whenever you restrict food intake, you're going to run up against your own biology," explains Dr. Sharma. "It doesn't matter what program you follow. As soon as your body senses that there are fewer calories going in than going out, it harnesses a whole array of defense mechanisms to fight that." When we're dieting, our bodies try to conserve energy, so our metabolism slows down, the result being that you have to eat even less to keep losing weight. That becomes an increasingly difficult project because our bodies also produce more of the hormones, such as ghrelin, that trigger hunger. There is even some evidence that the bacteria in our guts respond when we eat fewer calories, shifting their populations in ways that will send more hunger signals to our brains.

All these weight-loss deterrents are hardwired into our biology because maintaining our body's size and fat stores is essential to human survival. This makes sense when you consider how many millennia humans spent living in food-scarce situations. Today, your body doesn't know if your pre-diet weight was too high; it's programmed, through eons of evolution, to protect that "set point" at any cost. "It might be six months later, it might be five years later," says Sharma. "Your body will continue to try to get the weight back. And eventually, it wins."

Almost nobody understands what a huge role biology plays in weight management. The set-point theory has gotten plenty of attention in the popular press, but even if it hadn't, most of us have gone on diets, so we know firsthand how difficult they are to maintain. But we don't accept biology as the explanation because we're

so convinced that weight is about willpower. When we start feeling hungrier, or thinking more obsessively about food, we assume
that it's our own human frailty at work, yet again. Indeed, 75 percent
of Americans cite willpower as one of their biggest barriers to
weight loss—and almost 60 percent of Americans believe that it's
a person's individual responsibility to lose weight, according to the
University of Chicago study. And most of us are sure diet and exercise is the best way to do that, even though half of people who say
they've tried to lose weight this way report doing so five or more
times over the course of their life. (A fifth of obese people say
they've tried and failed at diets more than twenty times.) "Most
people think there are good people and bad people, and good
people are the ones who have willpower and can fight their body's
urges to eat," explains Kaplan. "I'd argue that nobody can do that
except those with anorexia nervosa."

This, then, is why the hype around bariatric surgery, which
researchers now believe achieves its magical 25 percent to 35 percent
weight loss by permanently lowering your body's set point.
Although it has been long assumed that the procedure was successful because smaller stomachs hold less food, studies by Kaplan
and others have demonstrated that reshaping the stomach actually
reprograms all those fat-defending mechanisms. Bacterial flora in
the stomach take on different roles. The gut sends different hormonal signals to your brain, pancreas, and liver, which affect how
you experience hunger and fullness. For most adults, a daily diet
of 1400 calories would feel punitive; we'd battle constant hunger
as our bodies tried to persuade us to eat more. "But after surgery,
someone can easily live off fourteen hundred calories per day and
not feel hungry," explains Sharma. "You aren't starving, because
your body isn't trying to get those calories back." It doesn't need

them, because it's not trying to get you back to your original set point of say, 300 pounds. It's picked a new set point—say, 195 pounds—and is perfectly happy to leave you at this new, lower weight for now.

The surgery also appears to change your body's metabolic processes and even your taste buds, probably as additional ways of reducing hunger. "After surgery, people can detect sugar in water at much lower thresholds than before," says Randy Seeley, Ph.D., a professor of surgery at the University of Michigan School of Public Health. Patients also gravitate toward eating less calorically dense foods and smaller but more frequent meals. "Surgeons often think the reason their patients start eating salads instead of Big Macs is because they've had a serious conversation about the importance of making these changes," Seeley notes. "But these people have been hearing those magical instructions their entire lives. Now they can actually follow the advice because their preferences have changed."

Seeley first demonstrated these changes through animal studies. "It's easier to show this in rats because they're not making food choices based on some idea about what weight they should be," he notes. "But they nevertheless reduce their food intake for several weeks, then go back to eating normally and maintain a fifty percent reduction in body fat for the rest of their lives." Unlike many humans, rats ultimately seem to be able to eat exactly the same amount of food as they did before the surgery—but only once they've lost the weight. When Seeley deliberately starved his rats so they lost even more than half their body weight, then gave them unlimited access to food, the rats regained until they got to their new, lower post-surgery weight. "We've changed that set point," he says.

When I ask Seeley why rats lose more weight post-surgery than people do, he says he thinks it only looks that way because humans diet. "Most people have been trying to override their biology for years through dieting, so even though they're very overweight, they're already somewhat below their body's natural set point before they get to surgery," he explains. Ironically, many insurance companies also require patients to achieve some degree of weight loss through medication or lifestyle changes before they'll approve the procedure. That sets patients up to feel like failures when the post-surgery weight loss isn't as dramatic as they hoped, because they aren't giving themselves credit for what they managed to lose in a few weeks or months of crash-dieting beforehand. They want to have lost that and then shed an additional hundred pounds. "Of course, without the surgery, they would have regained all of that and more," Seeley says.

Seeley's research so underscores the power of these biological changes that he argues that surgery patients don't need to be put on prescribed diets that restrict their caloric intake. "We do need to educate them about taking extra vitamins and getting enough protein now that they're eating fewer calories," he explains. "But my intuition is that adding calorie restriction on top of the surgery will be no more effective." It could even be counterproductive. Surgery patients who end up restricting calories so dramatically— "overshooting," as Ramona advised—that they end up below their body's new set point will find themselves caught in the same old cycle of weight loss and gain, albeit with lower numbers. Do that often enough and you can start to push your set point higher again, because we tend to gain back a little more each time. "And now it looks like the surgery has failed," says Seeley. Or that the patient has failed at surgery.

Three days after being discharged from her gastric sleeve surgery in July 2015, Rachel Adkins drank a doctor-prescribed protein shake and immediately began vomiting. Within hours, she had a fever of 103. Her husband drove her to the emergency room near their house in Fishkill, New York, where a CT scan revealed that a leak had sprung in her new, banana-shaped stomach. During her initial recovery in the hospital, doctors had tested the staples holding Rachel's stomach shut by having her drink a barium solution, and then watching how the liquid moved through her digestive tract on ultrasound. Everything seemed fine. But now one of those staples had popped out.

The protein shake, along with anything else in her stomach, began pooling into her abdominal cavity. Rachel developed sepsis, a body-wide infection that can lead to tissue damage, organ failure, and even death. She was admitted to the intensive care unit and put into a medically induced coma for two days while doctors worked to save her life. A week after she was discharged a second time, Rachel's left lung collapsed because of accumulating fluid, another post-operative complication that is risked with any kind of thoracic surgery. Over the next two months, Rachel underwent six additional surgeries and spent more nights in the hospital than at home. For the first seventeen days, she ate nothing at all, relying solely on IV nutrition. For another twenty days, she was fed adult formula through a gastrostomy tube, which surgeons implanted in her abdomen. "It's very strange to feel full when you're not putting anything in your mouth," Rachel reports, confirming something I always suspected during Violet's G-tube tenure.

Terrifying and life-threatening consequences like these have become more rare as bariatric surgery has advanced, but they still happen. Other complications that aren't life-threatening, but that are certainly life-hampering, are more common. Many patients develop nutrient deficiencies caused by the body's new inability to fully absorb what it needs from food. And 85 percent of people who have the gastric sleeve or Roux-en-Y procedures will experience "dumping syndrome," in which undigested food hits the small intestine too quickly and causes pain, bloating, vomiting, and diarrhea. And all of that is just what may happen in the first few months. "There are a huge number of things that nobody has figured out yet, like how do medications get metabolized by your body now?" Burgard points out. "What we really need to ask is, how is it going to be in someone's long-term health interest for them to essentially be malnourished for the rest of their lives?"

The answer to that question depends on whether you think weight is synonymous with health. To pretty much everyone in the bariatric-surgery community, it is, which means that achieving permanent weight loss trumps all other concerns. Just as when treating a child who doesn't eat, behavioral feeding therapists focus on getting calories in by any means necessary—no matter the risks to that patient's long-term relationship with food—bariatric surgeons think solely in terms of getting calories *out* of their patients' stomachs, because they are so sure that doing so will save their lives. Despite the complications she experienced, Rachel also frames her surgery as a life-saving intervention. At five feet, six inches tall, she weighed 437 pounds when she decided to have the procedure. By early 2015, she had developed type 2 diabetes, along with elevated blood pressure and cholesterol. She had also given up teaching dance, a lifelong passion, because she felt her size made it too difficult to be so physically active. And her surgeon told her afterward that once he opened her up, he could see that she was close to liver failure.

Not every obese person experiences these kinds of health issues. Sharma is the author of research showing that around one in five obese people have very few, if any, related health problems and that for people in the overweight range, the connection between weight and health is even less pronounced. Yet he is still quick to frame bariatric surgery as more about health than weight loss. "I send patients for surgery when they have severe obesity and a health problem for which weight loss is the best solution," Sharma explains. "Let's say they have really bad pain in their knees. To get out of that wheelchair, they have to lose some weight." As a moderate thinker on these questions, Sharma doesn't demand extreme weight loss: "They don't have to hit a BMI of 25. If you define success based on whether their health problems get better, we know that even just a five percent weight loss leads to very significant improvements." But he does emphasize that bariatric surgery has been linked to a 50 percent reduction in patients' risk of dying from heart disease as well as a 90 percent reduction in a patient's risk of dying from diabetes (though only about half of diabetic patients are able to come off medication entirely after bariatric surgery). The data on other conditions, like sleep apnea, is less clear; patients may be less dependent on respiratory assistance, but most will still need support.

To many doctors, minimizing the threat of these conditions through weight loss far outweighs the relatively smaller risk for severe surgical complications, and even the much larger chance that a patient is signing on for a lifetime of complex nutrition management and digestive issues. But there is another interpretation of the literature. In the late 1990s, while researching her doctoral dissertation on the relationship between health and weight, Linda Bacon, now an associate nutritionist at the University of California at Davis, was shocked to discover a large swath of research suggesting that being classified as overweight or obese on the body mass index was a deeply flawed indicator of one's health and longevity. "Every study I found suggested that the BMI cutoff points for overweight and obese should actually be raised; that we were putting too many people in those categories when their weight didn't actually correlate to much in the way of health problems," she says. Instead, in June 1998, the National Institutes of Health's

Obesity Task Force lowered the cutoff points. "Just like that, twenty-nine million Americans who had gone to bed with normal, healthy bodies woke up the next day and were fat," Bacon explains. "The task force had looked at all of the same evidence as me and essentially thrown out the data." Bacon draws a line between that decision and the marketing of two weight-loss drugs by major pharmaceutical companies. "If you make more fat people, you have a bigger market."

But as she continued to study the issue, Bacon began to collect more evidence for her argument that it isn't fat cells that are killing us—it's a combination of several other factors. Unhealthy lifestyle choices are the most obvious, but Bacon also points to social inequalities. For example, rates of diabetes, asthma, and other conditions often linked to weight are also higher in neighborhoods like Sherita Mouzon's, where people struggle with systemic poverty and discrimination, and also have less access to safe outdoor spaces, walking paths, and grocery stores. And Bacon argues that doctors fail to take into account the detrimental health effects of living with weight stigma, which has been well documented as a risk factor for depression and low self-esteem. Studies have also found that doctors and other healthcare providers spend less time with obese patients, offer less education, and even admit to liking their fat patients less. Such distaste is certainly not lost on obese patients, who report seeking medical care less often because they expect to be shamed for their weight—and who may well end up sicker as a result. In 2008, Bacon made her case in a book called *Health at Every Size: The Surprising Truth About Your Weight*, and the HAES movement entered public consciousness in a much bigger way.

If you're now thinking, "Okay, sure, walking paths are great, but we still need to lose the weight," consider Bacon's favorite analogy: Yellow teeth are common among lung cancer patients—but that doesn't mean a patient's teeth *caused* her lung cancer. It means that both are things that can happen when you smoke cigarettes. And correlation is not causation. Excess body weight, Bacon argues, hasn't been shown to cause the litany of health problems we associate with it—all we know from decades of epidemiological research is that higher body weight often coexists with diabetes, heart disease, sleep apnea, and so on. But just as having your teeth bleached won't improve your lung health, significant weight loss isn't the most logical tool if you want to lower your blood sugar or blood pressure—mostly because such weight loss isn't attainable or sustainable for the vast majority of us; remember those 84 percent of Weight Watchers clients, returning to the fold every few years.

Moderate weight loss does correlate with improvements in many of these conditions, but again, it's difficult to parse out whether those benefits accrue specifically from having a smaller body, or from adopting certain lifestyle habits—less soda, more sleep, less television, more walking—that sometimes also lead you to drop a jean size. "I'm not saying that everybody who is fat is healthy. I'm not saying that everybody is at an ideal weight. I'm not dismissing the relationship between weight and health," Bacon says. "What I am saying is that the relationship between weight and health has been wildly exaggerated. And when we focus exclusively on fat, we miss opportunities to make changes that have been proven to be successful."

Bacon's argument has resonated with me since I first read her book almost a decade ago, but I've been trying to choose a clear winner in the health/weight debate for just about as long and I'll admit, it often feels like an impossible task. Both sides are impeccably footnoted, with decades of studies; for every published position paper claiming that weight really does trigger health problems, there is another emphasizing that correlation is not causation—and back and forth the arguments fly. There are smart, good, caring people working on both sides of this aisle. But I will note that while there is significant money to be made in weight loss (whether you're a drug manufacturer, a diet coach, or yes, a bariatric surgeon), there is far less profit in telling people to stop trying to change their bodies. And I find it revealing that the purported health benefits of weight loss are not the bariatric-surgery industry's primary marketing tool. Visit the website of any bariatric surgeon, or jump into an online support forum, and what you'll mostly find are sad Before and triumphant After photos. "The right reason for any patient to seek surgery is better long-term health outcomes and not 'to fit into their old clothes,'" says Seeley. "But are some patients more motivated by the weight loss? Sure."

It's not just that "some patients" are more motivated by weight loss. *Everyone* is motivated by the promise of weight loss, whether we weigh 400 pounds or 140. When asked about their motivation, 93 percent of dieters said "overall health" was a major or minor driver—but nearly as many (89 percent) said they were driven by a desire to improve their appearance or "how my clothes fit." In other words, many of us aren't trying to make ourselves smaller just because we're passionate about preventing diabetes. We care about being thinner because our culture prefers smaller bodies and well-controlled appetites to big, unruly ones. And bariatric surgeons, obesity researchers, and other healthcare professionals are not exempt from this bias—in fact, research suggests that as a group, they display even higher levels of it than laypeople do.

Rachel's first real meal was a small plate of finely chopped chicken, served as a test run near the end of her hospital stay. Rachel's father sat with her while she ate it, trying to be positive. "He kept saying 'That looks delicious!'" she recalls. "And the weird thing is, it was. I mean, in retrospect, it was probably so gross. But I hadn't eaten food in so long; I hadn't had any kind of texture in my mouth. So that texture felt amazing." She ate a few spoonfuls and immediately felt full. Such a small portion would never have satisfied Rachel before her surgery, when she regularly consumed upward of six thousand calories per day. "I was a secret car eater," she says when we meet for lunch in March 2017. "I'd try not to eat all day, then hit the drive-through and get three cheeseburgers. It was super shameful."

The plate of chicken was actually the start of Rachel's second time learning to eat again after weight-loss surgery. She had a gastric band placed in 2010, but had to have it removed four years later because it caused vomiting whenever she ate something even a little bit tough to digest, like meat or even crackers. "If I didn't chew my food thoroughly enough, or if I ate something kind of dry, I'd feel it getting stuck in the top of my stomach and then it would come right back up," she recalls. "What did stay down was milkshakes, pasta, all the bad food that's easy to slide through." Rachel weighed 335 pounds before the lap band surgery and lost 110 pounds within the first year. But she ultimately regained it all, and then kept gaining after the band came off. "Once the restriction of the band wasn't there anymore, I just fell back into eating whatever, whenever I wanted," she explains. "I don't think my body or my mind was ready to deal with my issues."

Because gastric banding is designed to be temporary, research suggests that patients don't see the same kind of profound biological shifts in taste and metabolism as have been documented with permanent bariatric procedures. But even with permanent gastric-bypass procedures, the pronounced changes in food preferences, for example, seem to dissipate for many patients as they adjust to life at their new set points. Lab rodents appear to adapt to this and maintain their new set point despite being able to eat more food. But the return of a larger appetite is much more fraught for human patients. If you frame weight loss as a matter of willpower, suddenly feeling hungrier when you aren't yet "thin enough" feels like you aren't trying enough. You have to reckon once again with craving "all the bad food," as Rachel puts it; with once again wanting to eat the second half of the protein bar that Ramona is so convinced they shouldn't need. Old habits, like binge eating, may return.

Marci Evans, a dietitian with a private practice in Boston, treats people struggling with disordered eating post-surgery, and wonders how much a surgeon's technique can shape both the intensity and the duration of any resulting changes in hunger and fullness cues. "Anecdotally, I've seen patients from a surgeon who I know to be rather conservative do much better afterwards than folks who have had their stomach so dramatically altered that they're going to have to figure out how to live on five hundred calories per day and a ton of supplements," she says. She points to one client who had one of those more conservative surgeries, meaning the surgeon left more of his original stomach intact. He had also done a lot of work on his emotional eating issues prior to his surgery. Afterward, he expressed relief that his hunger now seemed better regulated and closer to "normal." He was able to eat what Evans considered to be an adequate diet with plenty of variety. But she says he's the exception to the rule. "My clients are often very unsettled by the way their body is now responding to food," she explains. "Because psychologically, it's not like they got a lobotomy. They still have that drive to eat." We're not rats, isolated in our cages. Any changes to our biology have to interact with our psychological and cultural understanding of food. And nobody knows yet precisely how one shapes the other.

As a social worker in the outpatient psychiatry program at Massachusetts General Hospital, Lisa DuBreuil has worked with people

struggling with eating disorders and drug or alcohol addictions for
the past fourteen years. About nine years ago, she and her colleagues began noticing a new pattern of patients: Adults in their
thirties, forties, or fifties who were suddenly developing a substance
abuse problem or an eating disorder for the first time in their lives.
And these newly struggling patients had something in common.
They had all undergone some form of weight-loss surgery. Seeley
calls the increased risk for alcoholism "small but real," and when
DuBreuil dug into the literature, she found that as many as
28 percent of patients develop a problem with alcohol. For two
thirds of them, this is a new addiction. Nobody's sure why, but
research that tracked people after Roux-en-Y bypass surgery found
that their bodies were more sensitive to alcohol post-surgery. "Drink
for drink, their brains are exposed to more alcohol now" because
they metabolize it differently, DuBreuil explains. "And that
increases your risk for developing a problem."

This change in alcohol processing may be related to the other
biological shifts that occur with surgery. Another theory is that
during the initial post-operative period, when food becomes less
rewarding, people may be more inclined to seek out alternatives,
especially if they once relied on food as a coping strategy, or as a
way of connecting emotionally and socially with others. DuBreuil
frames the issue a little differently: "Many of the people I work
with struggle with the fact that they don't get hungry anymore,"
she explains. "But they're also carrying around decades of cultural
experiences and psychological attachments to food. So yeah, maybe
their preferences have changed and they don't like cookies anymore.
But now they have to contend with the fact that they're from a big
Italian family and everyone is supposed to adore Nana's cookies,
and how do they manage not wanting to eat them anymore?"

Some experts explain the rise in alcoholism after surgery as a
kind of "addiction transference." Rachel now considers herself

a recovering food addict and looks back on her months of life-threatening surgical complications as a kind of hard-core detox program. "You learn really quickly that you don't have to eat to survive," she tells me. At first, I'm confused: We do, of course, need to eat to survive. But for Rachel, "eating" had always meant binge-ing; she only felt satisfied by a meal when she'd eaten a huge quantity of food and reached the point where she couldn't fit another bite in her body. "Of course, twenty minutes later, I would feel awful. But I had this mind-set that if I didn't eat a ton of cheeseburgers or other shitty foods, I would die," she explains. "Now I know I don't need ten thousand calories in a day. I can live on much smaller portions and feel fine."

And, just as Seeley's research suggests, she doesn't even crave the same kinds of food as she did before the surgery. For most of her life, Rachel's favorite foods were pasta, bread, and dessert. She's the oldest of five children, and all her family's celebrations revolved around that kind of comfort food. Like Gina, she spent years cycling between intensive dieting efforts—Rachel's list includes Weight Watchers, Atkins, SlimFast, and "those diet pills that make you feel like you're on speed"—and periods of intense binge eating. Every diet failed after two or three months because the deprivation was unsustainable. She couldn't replace the foods she craved with healthier alternatives and feel the same kind of sat-isfaction. "My taste buds are one hundred percent different now," she says. "I was never a vegetable eater and now I really do like them."

It's impossible to know if the shift has occurred because Rachel has finally internalized the importance of vegetables for her health, or because her biological food cravings have been reprogrammed. But whether that means that she was previously addicted to the pasta and desserts she now avoids is a controversial question in obe-sity research. When rats at the Monell Chemical Senses Center in

Philadelphia were given unlimited access to chow that had been flavored with a noncaloric sweetener or a noncaloric fatty flavor, they didn't gain any more weight than when they were given unlimited access to their regular, unflavored chow. This finding prompted the study's lead author, Michael Tordoff, Ph.D., to conclude that it isn't taste that drives us to overeat certain foods. He hasn't yet attempted the study with human subjects, but suspects the results would be similar. "How a food tastes might trigger you to eat more the first time you have it, but it doesn't compel you to eat it endlessly," he says. "The evidence that food can be a true addiction doesn't really hold up." Seeley agrees: "If you are hungry, you will obsess about obtaining food. That is how the system is supposed to work. I don't think that counts as addiction."

The food-addiction theory may also be another way of stigmatizing the choices we make around food. "The thing is, we're all food addicts," says DuBreuil. "We're neurologically wired to enjoy food and to want to eat when we see others eat. We're supposed to connect food with love and caring." After all, for most humans, the first experience of eating is intertwined with love, when our mothers hold us and feed us with their own bodies. "Yet our culture pathologizes the idea that food is love," she notes. "We're just supposed to eat to survive." To derive too much comfort, too much pleasure from food is interpreted as a sign of addiction, but DuBreuil argues that the behavior even of eating-disorder patients desperate to binge is quite different from the life-destroying choices her drug- or alcohol-addicted clients sometimes make to obtain their highs. And perhaps if we stopped shaming people for loving to eat, they could love food in a less complicated and frenetic way.

If food addiction is real, the hardest thing about it must be that going cold turkey is not an option because we all have to eat food and we have to do it many times a day. After her "detox," Rachel remembers feeling terrified to eat a two-ounce serving of mashed potatoes or yogurt once she was finally cleared to eat solid foods again after the chopped chicken. "I didn't want to put anything in my mouth that could rip that hole open again," she says. She also didn't trust herself not to slip back into old habits. Eating didn't feel safe. Three weeks after she was discharged, Rachel's parents took her out to dinner at one of their favorite restaurants. But instead of celebrating, Rachel sat in the booth and cried. "I was so scared to eat this turkey dinner," she says. "I didn't want to make myself sick. I didn't want to overeat. I was scared to put anything in my mouth." DuBreuil has heard similar stories from her patients; she believes that the stringent framework for eating that patients are subject to after surgery further reinforces the feeling that they can't trust themselves around food. She also hears how they grieve for the loss of old mealtime experiences. "Sharing meals is probably our best way to bond and connect, other than sex," she notes. "It doesn't make you an addict to miss that. It makes you human."

Sometime after Gina decided to stop dieting, she began reading about the HAES movement and trying to figure out whether she could find a way to accept being fat. It was true that she didn't have any of the traditional obesity-related health problems. And she understood that eating well and exercising made her feel better regardless of what happened with her weight. She wanted to feel at peace with her body, but she was also still using food to deal with every kind of emotion, especially negative ones. She weighed 250 pounds when she and Nick got married in 2010. They started trying to have a baby a few months later, but it didn't happen. Every month, Gina faced up to that disappointment, and every month, she coped by eating. She also stopped exercising regularly because she had a long commute and a stressful job. On her lunch breaks, she would drive to the nearest McDonald's and order a

Quarter Pounder with cheese and fries, plus a four-piece order of chicken nuggets with sweet-and-sour sauce. Then she would eat it all while sitting in her car, listening to NPR. "It was pretty luxurious," she says now. She felt better—comforted, relaxed, and numbed—in the moment. Until the guilt hit.

Gina had always suspected she might have a binge-eating disorder, but whenever she brought it up, her mother would say, "No, you're fat. If you had an eating disorder, you'd be thin." But sometime in 2011, as their efforts to conceive became more fraught, Nick started to notice Gina's tendency to cope through food. "Nick was like, 'Hey, your eating is fairly disordered. I noticed you bought that chocolate cake and ate the whole thing when you got upset,'" she recalls. "And it was okay, because it was a husband talking to his wife." Gina began to realize that eating the whole cake didn't make her feel better. "It actually makes you feel like shit."

Over the next few years, Gina did a lot of work on her emotional eating. She stopped driving to McDonald's for lunch and started going for walks, and then getting a salad or sandwich at Panera. She started exercising after work—not to lose weight, but because she started to see how exercise could provide an outlet for all those negative emotions. "I'm just a much nicer person when I'm exercising," she says. One day she found that she could eat a slice of cake—but just a slice. Gina never sought professional treatment for her binge eating; years of relying on misguided diet "experts" left her determined to do this herself. It wasn't always a straightforward process. And it didn't correspond to getting thin; Gina's weight continued to climb, hitting 280 and then 325. She still wasn't getting pregnant and the infertility anxiety made everything harder. The couple met with two infertility specialists and they both said the same thing: Obesity was probably causing all of their problems, but unless Gina lost weight, she wouldn't be a good candidate for in vitro fertilization. They both recommended bariatric surgery.

"I kind of hit rock bottom, when the second doctor told us that," Gina says. "I always thought if I'm so fat that I need surgery, that means I've failed as a person." But then she realized that it wouldn't really be about the weight loss. "I was at the point where I really would have cut off an arm to have a baby," she explains. "And once I got there, it was like, 'Okay, why not cut off part of my stomach?'" The irony, of course, was that by the time Gina made the decision, she hadn't had a binge-eating episode in almost a year. She really did feel good about her body, about being a fat person who defied stereotypes by exercising and eating well. And now she felt as if she was about to undermine all of that, by trying—in the most dramatic way possible—to make herself permanently thin. "I felt like a fraud," she says. "But I really wasn't doing this to be a certain size."

In a roundabout way, Gina's decision to abandon the elusive goal of weight loss may be why she has been so successful since having her surgery. Unlike many patients, she's not disappointed to find herself still obese. "A girl stood up at our surgery patient support group and said she's still waiting to get her bikini body," Gina tells me. "I think: 'You have a body. Put a bikini on it if you want to.'" In the first weeks after her operation, Gina was fascinated to realize how food would now work—that in the early months, two tablespoons of peanut butter would leave her full for several hours, and that even now, a year later, a hundred-calorie cup of yogurt does the same. But even though her overall caloric intake is less now, and she eats smaller, more frequent meals, Gina doesn't think of how she's eating as a diet, but rather, as what works best for her body. "I truly believe there are no good foods or bad foods," she says. "Nothing is off limits. So if I want some ice cream, I have it. The difference is now I don't eat my feelings."

I see this in action when we're at lunch. Gina orders corned beef sliders and a salad off the appetizer menu. But she has a bite of Nick's fried chicken and later, a few forkfuls of my Oreo cheese-cake for dessert. If I didn't know she'd had the surgery, I'd think nothing of her choices. In contrast, when I go to lunch with Rachel almost two years after her surgery, she orders only a side of spinach artichoke dip. It comes in a small ramekin surrounded by pita chips; she ignores the chips and eats only a few spoonfuls of the dip before stopping. I'm acutely aware of how much longer it takes me to eat my Chinese chicken salad, of how much food is left on her plate. Often, Rachel tells me, she doesn't even order for herself in restaurants anymore, but just picks off her husband's plate. At home, she weighs and measures everything to stay on top of her portion sizes. Holidays, work parties, and other social gatherings that revolve around food are stressful in a way they never were before, because Rachel knows she'll not only have to navigate how to eat, but also the curious looks and "Oh you're so good!" commentary from people who can't believe how little she's eating.

DuBreuil says that when she first started working with post-surgery patients, she had to recalibrate her responses to habits like not ordering in restaurants, weighing portions, or chewing every bite twenty times, all of which would be listed as classic examples in any textbook on disordered eating. "But for these folks, it's the new normal," she says. And in the right context, such habits can be a form of self-care: "If you're doing these things because you're listening to your body, that's different from doing them to get your body down to a certain size," DuBreuil says. The hard part is telling the difference. In their quest to reach a certain size, surgery patients have already agreed to drastically alter their anatomy. How do they now listen to their body after having cut away a part of it because a doctor told them it couldn't be trusted?

This is why many in the HAES community are so critical of

the procedure's risks that they're reluctant to endorse any version of it. When I emailed Linda Bacon to ask for her thoughts on weight-loss surgery, she wrote back that she wasn't interested in talking: "I'm just pained and disgusted by what goes on behind the scenes," she said, noting that few studies have been done by researchers without some bias or a conflict of interest. But DuBreuil and Evans, who both identify as HAES-oriented practitioners, feel strongly about the need to find common ground. "I think we're starting to realize that actually, it's the people who go for these surgeries who need the most support," says DuBreuil. "Compared to larger-bodied people who don't pursue weight-loss surgery, these folks are more likely to have diagnosable anxiety, depression, and other mood disorders. They're more distressed by living in their fat bodies."

Still, DuBreuil acknowledges that she'd rather see fewer patients resorting to surgery. The HAES approach to health is arguably kinder, gentler, and safer. After all, you get to keep your stomach and enjoy your food. But it's not necessarily easier. It means abandoning the notion that you'll someday be thin, or at least, significantly thinner than you are now. That's a difficult dream to give up on, especially for women who have been conditioned to want it since we were given our first Barbie dolls. And especially for obese people, who have to put up with a near-constant train of unsolicited advice and opinions about their size and their eating habits. When Gina was nineteen years old, she walked out of a gas station holding a bag of candy and a guy leaned out of his car to yell, "Fat bastard!" at her. "I actually wasn't totally ashamed of my body by that point," she recalls. "I knew I was fat, but I didn't realize I was heckled-by-strangers fat until that moment. Then I told my family and friends and they weren't all that surprised."

And street harassment isn't the only way fat people are stigmatized; many public health campaigns trade in various forms of weight stigma as a way of giving obese people wake-up calls. In 2012, Georgia received criticism for a campaign called Strong4Life that featured photos of fat children with slogans like "Fat kids become fat adults" and "It's hard to be a little girl when you're not." If the goal of such messaging is to persuade fat people that they should try hard not to be that way, Gina says, it sort of works. "That one comment at the gas station galvanized me into hyper-dieting mode for pretty much all of my twenties." Stop trying to lose weight, stop internalizing such messages, and you're always going to be the fat person daring to eat candy in public. Or interviewing for a job. Or getting your annual checkup and hearing that maybe you should think about losing a few pounds, no matter how good your health is right now.

It's understandable that many larger people would rather abandon the fight to make the world accept them at that size. And maybe Seeley's research will ultimately make it easier for them to do so. "My goal is actually to put the surgeons out of business," he tells me. The more he understands about how surgery works to change appetite, the closer he is to developing a drug that would do the same thing, without the same risks. "If scientists can come up with something that lets people change their body size as safely as we can change our hair color, I'd actually be less concerned about it," DuBreuil says. "Because I believe everyone has the right to body autonomy." But she also wonders how that ability would further reinforce our cultural standards of beauty, already so often conflated with our definition of health.

For Rachel, nothing about either of her surgeries felt as easy as changing her hair color. And yet, when I ask whether she'd do it all over again, she doesn't hesitate: "In a second," she says. Rachel is still obese, but she has lost 180 pounds. She's teaching dance classes again. And whether because of the weight loss or her healthier eating habits and increased physical activity, she no longer needs diabetes medication. Her liver function is normal and she has seen all her other measures of health improve. "It's like I've had this whole awakening," she says. The specific eating regimen has become close to second nature. "I feel like I enjoy food in a different way now," she tells me over the half-eaten spinach dip. "I enjoy flavors. Before, it was just about quantity; I got pleasure in my body when I ate a huge amount whether it tasted good or not. Now I enjoy the taste even if it is just for three or four bites. And I feel good when I finish a meal and I didn't overeat. I know I'm not going to feel that shame and regret later." Rachel is reaching the end of the typical post-operative weight-loss trajectory; all the data suggests that she's likely at her nadir and can expect to regain some of what she's shed. Still, she says, she hopes to lose more weight. "I just want to feel healthy," she tells me. Somehow, despite everything that has changed for her, that label continues to feel out of reach.

Seven

Learning to Eat

On a bright, sunny morning in May 2016, Dan and I stand in a hospital's parking lot and decide to put Violet's feeding tube back in.

Inside the big white building, our child, who is now almost three years old, who adores ladybugs and has recently learned to spell her own name, lies semiconscious in the pediatric intensive care unit. There are three chest tubes draining excess fluid out of her abdomen. She is hooked up to oxygen and several IVs. The one in her neck is connected to a dialysis machine, which makes a terrifying sound we call the zombie apocalypse alarm every time she turns her head. Violet is in acute kidney failure, and although we don't know it yet, a softball-sized hematoma is collapsing her right lung. All this is the fallout from her third open-heart surgery, a series of complications that were both expected and wildly unpredictable.

When we first learned that Violet would need three heart surgeries, I couldn't imagine anything being harder than that first one, when she was so tiny, when we were such new parents and it was all so shocking and surreal. "The last one will be the hardest," my brother-in-law told me then; he was already a father and he knew about toddlers. "She'll be all about Mommy at that age." He was right. But it's not just harder because Violet needs us more now; because she can reach for us from her hospital bed and demand to be held. It's also harder because of how much more we need her. Because, when she is sick enough that she stops demanding things, I can sit next to her all day and still miss her so much.

Now, when Violet is awake, she stares at us with big, glassy eyes. Mostly, she sleeps. Our family visits in shifts, crowding into extra chairs around the room in what feels entirely too much like a vigil. I sit beside Violet and read long chapters from *Winnie-the-Pooh*, hoping my voice can comfort her, since I can't hold her. She has not eaten since the night before her surgery, eleven days ago.

For most of that time, I've come back to wondering whether she's hungry in that strange, idle way I remember from the first hospital stay when Violet was a newborn. There are so many more pressing concerns, and yet, of course, I keep thinking that I should feed my child. She eats now. It's been eighteen months since I injected her last tube-fed meal, over a year since she started drinking enough milk and water that we could stop using her tube for hydration, and six months since we had her gastric feeding tube removed. In our surgeon's office, Violet yelped as a physician assistant slid the tube out of its stoma. Then she readily accepted a lollipop and patted her bandaged stomach with pride. "We took my button out of my tummy!" she told us.

The stoma where her feeding tube used to live has now closed over into a kind of second belly button. Of all her scars, it's the one Violet is most aware of; she often prods it and talks about her "food button." But I don't think she understands that there was ever a time when she didn't eat. Her days revolve around Cheerios, blueberries, pasta, eggs. She even eats kale, if I stir in enough cheese. The night before her surgery, we went out for Mexican and she inhaled a cheese quesadilla with salsa and guacamole. We've been in a honeymoon period with food, where she is delighted and intrigued by new flavors. That's all over now.

The PICU doctors begin suggesting that we place a nasogastric feeding tube on day six of this stay; in fact, a resident notes at rounds one morning that the only reason one hadn't been placed was "parental objection." On day six, we refuse, because when Violet is awake, she asks for water, juice, and ice cream continuously. I can't understand why we would thread a feeding tube down the throat of a child willing and able to eat. In pediatric intensive care settings, NG tubes are dropped daily and considered a fairly low-level intervention, but I remember pinning down our baby while she screamed and Dan threaded the tube into her nose, then listened with a stethoscope to make sure it had landed in her stomach and not her lung. I remember a day when threading the tube into place took so many tries that Dan had to leave the house afterward, to go run a steep, rocky trail through the woods as fast as he could. I started the next tube feed and tried to hold an inconsolable Violet. She sobbed and coughed as formula ran through the tube. Minutes into the feed, she began to choke, turning purple. Then she vomited so violently the tube came back up along with all of the formula. Dan came home and I had to tell him that we'd lost the whole thing. Shaken and sweaty, we started the process all over again.

The residents who keep proposing that we place a new feeding tube have never lived with one. Even the senior staff, who have seen hundreds of patients with them, don't know what we know. And it's not just our own traumatic history. We have also seen how the attitudes of medical professionals can change once a child has been labeled as "NPO," that heartbreaking medical acronym meaning "nothing by mouth." Nutrition becomes a prescription; a series of metrics for doctors to experiment and tinker with. Unless we press to run the feeds ourselves, nurses connect the feeding pump and hang the bags of formula, mixed up in some anonymous kitchen next to the hospital pharmacy, by a person I've never seen. And tube-fed children are treated as less able. Nobody trusts them to know whether they're hungry or full. I can't lose Violet's status as an eater. I can't go back to that gastroenterologist mapping out my child's lifetime on formula, thinking only about preventing clogs in the tube, like a plumber who has just snaked a drain. I have to accept that I can't hold my daughter when she asks me to, that she can't go home and sleep in her "other bed," as she now calls it. I will not let her lose this last bit of agency.

But by day ten, Violet has stopped asking for food. She manages a few sips of water or Ensure, and then turns away from the cup just as she once swiveled away from a bottle and my breast. Violet's eyes are sunken, her limbs birdlike. Her body is running through her fat stores and beginning to steal calories from her muscles. She is too sick to eat. And so we sit at a big round table in the center of the PICU with Violet's attending physician, dietitian, and pharmacist and listen to them make their case. They have been providing some basic nutrients intravenously, but you can only put so many tubes into a child's veins and now we're using that one to run Violet's dialysis machine. And Violet needs nutrition to support everything her body is trying to do. It all makes sense medically.

I try to explain where we've been with feeding tubes, why this might not fix everything. I tell them how Violet stopped eating completely on the NG tube as a baby; if our goal is to get her eating again, any day spent on a feeding tube will contradict that. The doctor tells a story of another patient, severely disabled and nonverbal, whose mother fought to let him eat before surgery despite anesthesiology orders, because it was so important to her to be able to feed him. It was her only way to connect with her son. "Feeding is really important to moms," he says, to me, but also to the rest of the table, as if he needs to translate for them why I care so much. "It's very emotional for you." This is supposed to sound compassionate; I'm supposed to understand that he's trying to think about our big picture, to still see Violet and me as people beyond his prescription pad. And I know this man: He admitted Violet to the PICU twice under emergency conditions; he has saved her with chest drains and breathing tubes. He is compassionate. But our past is no longer relevant. For Violet to eat again may be my goal, but it is not, and never has been, theirs. "If she were my child, I would have dropped the tube a week ago," the dietitian says, looking straight at me with sober eyes. I hear what she isn't saying. *You're starving your daughter.* It hangs in the air between us. Dan ends the meeting. He walks me out to the parking lot, where I cry hot, angry tears. "I don't think we have another choice," he says.

I know he's right when Violet barely fights the nurse who carefully threads a new NG tube down her nose and throat, and into her stomach. I worried that she would somehow remember this sensation, even though she hasn't experienced it since she was five months old, but if it does trigger some primordial muscle memory, that is not apparent from her reaction. Now it's just one more tube among many. Our nurse begins a slow drip of food. I have won one battle—the dietitian agreed to buy a blender and feed Violet a mix of milk and baby food, instead of pediatric formula. I negotiated this by describing Violet's vomiting history; wouldn't it make more sense to give her foods we know she can tolerate, instead of experimenting with formulas full of unknown variables? I can't tell whether the dietitian agrees because she is actually swayed by my logic, or whether I'm just being placated. I don't even fully believe that tube-feeding food will make a difference. I allow myself to feel triumphant anyway.

We start the new feeding tube on a Friday afternoon. Early on Saturday morning, Violet throws up and pulls out the first tube. The nurse drops a second one; twelve hours later, Violet does the same thing. My stepmother is at Violet's bedside that night so Dan and I can sleep. She tells the PICU staff that they can't place a third tube in twenty-four hours without parental consent. The doctors agree to leave the tube out overnight so Violet can sleep and to reassess at rounds the next morning. It's the weekend; the dietitian is not there to be angry about it. And the staff who are on duty are more concerned with Violet's finicky dialysis machine and its never-ending zombie apocalypse alarm. On Sunday morning, somehow, we skirt around the discussion of replacing the NG tube. At lunch, Dan climbs into bed next to Violet and begins to eat a plate of hospital-made penne doused with a watery tomato sauce. She watches him for a few minutes, thinking. "That's *my* pasta," she says. And eats fifteen pieces.

My best medical explanation for that small miracle is this: Despite zombie alarms, the dialysis began to work, and Violet's appetite returned as her kidney function crept up. The tiny bit of nutrition she received through the tube before she began vomiting certainly also helped. We were mired in a catch-22: nutrition would promote healing, but Violet first had to heal enough to be able to eat. Used judiciously, a feeding tube can be the right tool to help a patient through that phase. But only if medical professionals remember that placing one doesn't render a child less capable; if they understand that although it may help resolve the immediate crisis, it also places a new roadblock between the child and regular eating—which is to say, between them and their regular life.

But mostly, having to put Violet back on a feeding tube confirmed what we already knew about our child (and indeed, most humans): She doesn't eat when she doesn't feel well or safe. In the PICU, you are neither. Over the next few weeks, we see this again and again, as Violet's appetite waxes and wanes with every new medical roadblock. Just forty-eight hours after her pasta lunch, she stops eating again as the hematoma flattens her lung, requiring emergency surgery. In the days following that procedure, her blood-glucose level spikes and she becomes wildly thirsty, begging for water, chocolate milk, and juice whenever she's awake. She grabs every cup and chugs down the liquid, drinking until she vomits. And so we must pace her, only allowing an ounce every fifteen minutes. It's just as brutal to refuse this intense need as it is to worry when she won't eat. One afternoon, Violet begs to drink almost every minute, coming up with increasingly creative ways to pose the request: "Just a little sip, Mama." "Just one more drink." "I need that water over there." "I want to hold your cup." And so on. We ban anyone else from drinking or eating near her, even if it's a food she doesn't usually like. ("I want *your* dinner," she tells me when I make the mistake of opening a container of sushi too close to her bed.) When I take a break to run to the bathroom, my stepfather reports that the incessant requests stop; Violet lies quietly until I return, and even allows Dan's mother to distract her with a picture book. Somehow, she knows that I am the primary person upon whom she should focus her efforts; that I am the keeper of all she needs and yet also the one now denying her. "Feeding is really important to moms," the PICU doctor said. "It's very emotional for you." He had no idea.

After a few days, Violet's glucose level evens out and she's less thirsty. Then she's not thirsty at all. Or hungry. We're right back to where we started, with a child who won't eat. And it's harder, this time, to parse out why, because there are too many good reasons. The doctors have prescribed several powerful mood stabilizers to help keep her calm, and they all have appetite-suppressing side effects. She's receiving intravenous nutrition again to prevent another glucose spike; doctors are mixed on how that can interfere with the desire to eat. A viscous fluid called chyle continues to accumulate around her lungs, as a result of her post-op complications. The chyle increases when Violet eats anything containing fat, so we switch to skim milk, fat-free pudding, SnackWell's, and other 1980s diet food. I make daily field trips to the nearby Shop-Rite and scour nutrition labels. What with the current vogue for low-carb and sugar-free food, fat is surprisingly tough to avoid. It doesn't matter. Violet hates everything I buy. Over the past weeks, we've taken her on and off NPO status so many times—pushed her to eat when she's not hungry, stopped her from eating when she is—I'm starting to wonder if she's shut down her eating instincts altogether.

At morning rounds, the team seems to grasp that there are multiple medical explanations for Violet's new loss of appetite, but they can't fix any of them without risking a new crisis. So instead, they switch to guilt trips. One doctor suggests that we tell Violet she can't leave her room to visit the fish tank in the hospital lobby unless she finishes her lunch. "Maybe she just needs to understand there are consequences to not eating," she says sternly. Others just look at us plaintively; every time I leave Violet's room, someone asks me hopefully, "Did she eat yet?" I try not to bring those questions into the room; we tell the nurses not to ask Violet directly about eating, but some do anyway. "Why won't you eat, honey? How about just three bites? Yum, that looks so good!" Violet is well enough to sit up now, to pretend to bake cupcakes with me on her tray table. She is perfectly aware of what everyone expects of her. Their expectations are not motivation enough, and as desperate as I am for her to eat, I am also just a little bit proud that she cannot be swayed. Maybe she is still listening to her body. It's just giving her the wrong message.

Finally, it is the weekend again, and we are once again free from the dietitian's daily scrutiny. We convince the most rebellious PICU doctor to drop some of Violet's medications and her intravenous nutrition. Within a few hours, she tells me she wants a turkey sandwich. I text Dan to get supplies. Violet and I laugh when we look out her hospital room window and see him sprinting across the parking lot, a bag of fat-free potato rolls held aloft in victory.

This time, Violet's appetite sticks. She'll eat bagels and toast with fat-free peanut butter, strawberries, Popsicles, and even the hospital's congealed fat-free mac and cheese. As we get through the rest of the summer, and two more surgeries, she seems to once again connect the dots between food and comfort. Food becomes a touchstone in the bewildering, liminal state of hospital life. When she can't get out of bed or fall asleep, she realizes, she can eat. The doctors are happy; she's getting the calories she needs. But I have a new anxiety: Our old standby, Ellyn Satter's Division of Responsibility in Feeding model, is abandoned. There are no meal schedules. We don't edit her food choices or refuse to cater to specific requests. If she says, "I'm hungry!" everyone leaps to respond and she can take her pick. Chocolate ice cream for breakfast. Bagels at three o'clock in the morning. Midnight Popsicles. I worry about the loss of structure, the lack of balanced nutrition—and about how many times we can destroy Violet's appetite and continue to expect it to come back, intact. That's the whole thing about our eating instincts: They work best when you interfere with them the least.

By fall, Violet is home, free of medical devices, and back in school. There are days when she still grinds her teeth at the sight of strangers. There are nights when she wakes up in tears from wordless nightmares. But she is also very much her old self, and she has largely returned to the normal eating habits of a three-year-old. She knows when she's hungry and when she's full. She gravitates toward foods that are easy to eat and quickly filling—pasta, chicken sausage, pancakes—because she's often more interested in playing than in sitting at the table. She'll also try the occasional bite of a vegetable as long as we don't make a fuss about whether she should. But she remains far more cautious about new foods than she was before the hospital stay. She has left the honeymoon phase of early eating behind and we are now firmly entrenched in toddler-neophobia territory. And the phrase "I'm hungry!" continues to have multiple meanings. She'll begin saying it as soon as she walks in the door at the end of the day, hanging on my leg as I try frantically to get dinner on the table. She'll demand a snack as soon as we clear away her plate, or in the car on the way to school when she has just devoured her breakfast. It is difficult, now, to know whether that kind of "I'm hungry!" truly means "I'm hungrier than usual today," or whether she learned, during those weeks in the hospital where we'd jump to grab another bagel, that those words have a special power.

When I started writing this book, I thought Satter's Division of Responsibility was the entire answer. "We just have to trust these kids," I thought. It worked so well when Violet was a baby. Now that she's older, I still see its wisdom, but I also know that it becomes increasingly harder to implement because there are so many more forces working to undermine a kid's ability to self-regulate with food. I have to find a different way to trust Violet and read her hunger cues, because "I'm hungry!" doesn't always mean that her stomach is empty. It also means "I'm bored," "I'm tired," "You're not paying attention to me," "I just saw those cookies on the counter and now I want one." Violet has learned to crave food for emotional reasons, to find comfort in it, which was all I wanted for her when she was a baby and again during these more recent hospital hunger strikes. But now, her emotional need for food feels more complicated. I'm worried she'll learn to turn to food out of boredom, when I'd rather she looked at books, drew a picture, or just dreamed inside her own mind. The comfort of food is often overshadowed by its power, especially the power of specific foods like ice cream, doughnuts, and chips. Keeping those treats neutral—not good, not bad, just food—is utterly impossible when nobody else around her is doing that. At Violet's fourth birthday party, I pass out slices of chocolate Oreo cake, and one little girl tells me, "It's not good to eat cake and candy every day." I am stumped, but offer, "Well, it's good to eat lots of different kinds of foods every day." She is not fooled. "I guess if by 'good' you mean 'tasty,'" she allows. "But cake and candy is *not* good for you."

Violet eats two slices of cake at her birthday party; the next day, she asks me what her "dinner treat" will be. A treat is, by definition, "an item that is out of the ordinary and gives great pleasure." It is not neutral. It is not something we eat purely for physical hunger. And her friend is right; a treat is not supposed to be an everyday thing.

But then I remember: Violet hasn't just now learned to eat for emotional reasons. Or rather, she has, but it's been part of her relearning. Because this too is part of the eating instinct. We're all born with a hunger, not just for food but also for comfort. To feed a baby without cuddling her—without even touching her—feels wrong. I know this because we had to feed Violet without cuddling her through most of her infancy. I felt the loss of that in a deep, visceral place; at times, my arms ached from not holding her in that way. I can't know what that loss felt like to Violet. It happened before she had language or memory. And after those first four weeks of her life, tube feeding was all she knew. Can a baby miss something barely remembered and impossible to articulate? I don't know. But I do know that the comfort–food connection came back. And it wasn't just a fringe benefit of learning to eat again—it was *how* Violet learned to eat again. She rediscovered her drive to eat when she understood that eating offered connection and pleasure. She eats less when it won't. The mere need for nutrition has never been motivation enough.

And I don't think it's enough for any of us. The goal of separating food and emotions is at the core of every diet plan. It's advice I have written into hundreds of women's magazine articles. What Violet experienced during her second round of eating struggles— the pressure to override her own appetite cues, the discovery that different foods hold different kinds of power—was just a fast-tracked version of what most of us learn, more gradually, throughout childhood and into our adult years. Our eating instincts are disrupted by modern diet culture, in which food is supposed to be fuel, not therapy. Just as the PICU doctors and dietitians think of nutrition as a prescription they can write and then tweak for optimal results, we're taught that a "healthy" relationship with food means that you only ever eat for sustenance. Enjoyment is allowed only when you're eating certain kinds of foods blessed with the right kind of packaging, or better yet, no packaging at all. Otherwise, we're supposed to ignore the sheer existence of food unless we're hungry, and then eat only what we need to feel full, but never a bite more. You shouldn't eat to combat depression, or stress, or just because something tastes good, if you are not also physically hungry. And yet—the physical sensation of hunger *is* emotional. Hunger triggers a huge range of feelings, depending on its severity—excitement, irritability, weepiness, confusion. And eating brings more: pleasure, contentment, satisfaction, bliss. We cannot separate these things. I'm not sure that we should try.

Karen is a fifty-one-year-old science writer who lives with her husband and children in California's Silicon Valley. She asks me to use just her first name, and to change the names of her husband and children, because her family story of food is so complicated. It stretches all the way back to Karen's first day of first grade, when she threw up a peanut butter sandwich in the school cafeteria. All the kids stared; her parents were called. To Karen, it wasn't the vomiting that was so traumatic; it was the fuss everyone made over it afterward. She knew she didn't like peanut butter, but nobody else seemed to understand how that could possibly be.

"In retrospect, I know my dad was allergic to peanuts and even though I don't go into shock when I eat them, I think I also have a little bit of that," says Karen now. "To this day, I will gag if I try to eat peanut butter." And peanut butter wasn't the only problem. The school cafeteria frightened her. It was loud and crowded, and she didn't want to eat anything offered there, or the peanut butter sandwiches that her mother kept packing for her, even after that first day.

Karen trained herself not to eat at school. Ever. By high school, she was signing up for an extra class so she could avoid being in the cafeteria altogether. "Sometimes I'd eat a yogurt in class or something; otherwise I'd just be really hungry when I got home from school," she recalls. When she was a kid, this didn't feel weird to her, though other children and teachers would sometimes make comments. It was just how she got through the day. Only in looking back does Karen realize how often she was tired and had trouble focusing in class. "And it's like, 'Oh, right, I didn't eat.'"

Karen grew up in a middle-class suburb of Buffalo, New York.

When she went to an Ivy League college, she felt like "a country bumpkin," compared to many of her classmates who had come from fancy boarding schools or from New York City. "That's when I realized, 'Oh, I've never been to a Chinese restaurant or really eaten any kind of ethnic food,'" she says. "That seemed normal for where I was from, but I was actually pretty sheltered." The dining halls at college were more pleasant than her public school cafeterias. Karen began to eat lunch, and then to eat a wider variety of foods. It was a conscious choice. She hated the feeling of going out with friends and not being able to eat anything on the menu. "I remember going to a sushi restaurant when I was twenty-two and feeling so awkward because I had nothing to eat," she recalls. "It seemed like something that might happen when you're seven, but not as an adult."

Karen was determined not to let it happen again, so she began to work at finding more foods she liked. By the time she turned thirty, she loved vegetarian sushi, as well as Chinese, Indian, and many other cuisines. "It was work to get over the initial hump, but once I found I liked some of these foods, it wasn't really work," she says. Certain textures and smells, like peanut butter and canned tuna fish, continue to freak her out. But for the most part, Karen has learned to enjoy food, though she'd much rather go out to eat than have to cook it herself.

But when she was thirty-one, Karen married Peter, whom she describes as a brilliant German engineer who speaks three languages. And almost every day, for the entire twenty years of their marriage, he has eaten four slices of whole-wheat bread with Nutella for breakfast, and four slices of whole-wheat bread with strawberry jam, plus a yogurt, for dinner. He will eat a wider variety of foods at lunch; Karen jokes that he has to keep his job at a major tech company because their chef-catered cafeteria is supplying most of his nutrition. But when they go to a friend's house for dinner, Peter will often eat in advance. When he makes dinner or packs lunches for the kids, they also get what the family call "Nutella boats." An entire kitchen cabinet is dedicated to storing those iconic brown jars.

Karen thinks her husband's eating habits stem from having been overweight as a child. "He's within the normal range of weight for his height now, but I think some of this rigidity has to do with wanting to maintain that." It may also be a response to how food was treated during his childhood in Germany. Peter's parents suffered years of food scarcity during World War II, when his father was held prisoner in a Siberian labor camp. "Whenever we go to visit, they push food on us in this almost pathological way," Karen says. "There's huge anxiety about not letting anything go to waste." From his parents, Peter absorbed the idea that you should buy the cheapest calories you can, in the biggest quantities possible. Although he eats such a limited diet himself, he regularly comes home with bags of dollar-store snack foods that he buys in bulk. "We're stuck at this intersection of his scarcity mind-set and our current time of food abundance," Karen says. Her husband is so rigid in his habits that he can ignore the sixteen boxes of cookies in the cupboard. "But that doesn't work so well for me or the kids." Peter's tendency to prioritize price over nutrition drives Karen crazy; she's embarrassed by the quantities of processed foods in their pantry. They live in California, after all—at the heart of the nation's alternative-food movement, where organic berries, kale, and a Vitamix for blending up your green juice are considered essential kitchen staples. Karen longs to be a part of that culture, even as she also feels oppressed by its high standards. "I find it hard to exist in today's judgmental food environment."

But Karen has long since given up trying to change her husband's eating habits. They have a good marriage and she accepts that this is part of the deal. "To be honest, as we get older, I have to acknowledge that his diet isn't giving him any health issues," she says, noting that she's constantly experimenting to find foods that don't trigger her acid reflux, while Peter suffers from no such ailments. "So I've had to reevaluate things. Maybe picky eaters do know what they want. Maybe we should be more accepting of their choices." Looking back at her own relationship with food, Karen wonders whether she would have become a more adventurous eater sooner if her childhood pickiness had seemed more acceptable, if she could have chosen what she wanted to eat for lunch instead of giving up on the meal altogether.

But she's nevertheless aware—and sometimes resentful—of how her husband's relationship with food has come to define eating for their entire family. By the time their daughter was a toddler, she would eat only beige foods, such as pasta and chicken nuggets. Karen was worried about how to handle Amanda's new pickiness, and her anxiety led to intense power struggles at the dinner table. It was hard to explain to a child why she should eat different foods every day when her father was eating the same exact thing night after night. Pickiness seemed contagious; pretty soon her son was also balking at many foods. Karen found Ellyn Satter and tried to absorb the Division of Responsibility ethos. But it didn't feel liberating to her. "There was too much shaming of parents who become short-order cooks for their families," she notes. For Karen, it was much easier to make what everyone wanted to eat than to endure the complaining and whining. She knew she was supposed to offer the children a new food twelve times and let them reject it twelve times before they'll eat it. "But I found it too traumatic to keep making food that would keep getting rejected," Karen says. "I think that's when I stopped reading parenting advice on the internet." Still, she felt judged for her decision daily. "I started to feel sort of ashamed to take my kids over to other people's houses, because it was always 'Will they eat this?' and usually they wouldn't."

So Karen soldiered on, making bunny-shaped Annie's macaroni and cheese for dinner and trying not to despair at her children's distaste for most other foods. "Amanda would have her bunny pasta and some milk and I'd rotate through her safe vegetables, which were broccoli, carrots, salad, or corn," Karen says. Then she would make a separate meal for herself, which her son might also pick at if he was bored with bunny pasta. Peter, of course, made his toast. And then, finally, after five years, Amanda decided to have something else for dinner. "I think she got tired of bunny pasta," Karen says. "It just kind of normalized on its own." Amanda is now seventeen and hasn't eaten bunny pasta in ten years, though she sometimes enjoys other kinds of mac and cheese as a comfort food. But she also eats Chinese food, pizza, vegetarian sushi, and many other foods that Karen says she herself wouldn't have touched as a kid.

Still, meals remain a struggle. Karen doesn't enjoy cooking and never wanted to be the primary cook in her household. "I have other things to do and I just think it's better when everyone contributes," she says. Peter was willing—but only if everyone was happy to eat Nutella or jam on toast. "That's not how I wanted to raise my kids," Karen says. So she has a rotation of a week or two of meals that she and her kids will all eat. She writes the menus on a board so everyone knows what to expect, and she handles all the shopping, cooking, and cleaning up. "It's a lot of work, especially for someone who is not a particularly joyful or creative cook," she says. "If I could, I would just outsource the whole thing." She especially hates chopping; healthy-meal-prep delivery services like Blue Apron and Sun Basket have become trendy in Silicon Valley, and Karen tried one, but canceled it after a week. "It was nice having someone else pick the recipe and supply all the ingredients, but I still had to chop and chop and chop," she says. "It took twice as long as it should have, to make sweet potato fries."

Karen doesn't think she'd push herself so hard to cook every night if "made from scratch" wasn't so revered in our food culture right now. "That whole thing of, 'It must be better if its homemade'—that's not true for me; most stuff I buy is better than what I make at home," she says. "I would just as soon buy it from someone else who knows what they're doing." If they order Chinese takeout or hit up the Old Country Buffet, everyone in her family can find something they like without her playing short-order cook. But then she fails on the nutrition front. "Either way, I can't win," she says.

There are some extreme elements in the way Karen's family manages food; I met her when I was researching adults with intensive picky eating, and certainly, Peter's Nutella habit might qualify him as at least a borderline ARFID case except for the fact that he's not troubled by his preferences. But look in the kitchen cupboards of most American households and you are likely to find odd combinations of ingredients or bulk snack-food stashes that have little to do with nutrition and everything to do with childhood, memory, habit. The most important part of Karen's story isn't the family's idiosyncratic food preferences or the failure of Division of Responsibility to resolve them. In fact, I think Karen's low-key approach to her daughter's picky-eating phase was not nearly as contrary to Satter's philosophy as she believes. Pairing safe foods with less accepted ones—as long as everything is offered in a no-pressure way—is a strategy endorsed by many child-led feeding therapists because it gives the child a sense of comfort, as well as the freedom to explore. And in Amanda's case, it seems to have paid off, albeit much more slowly than Karen would have liked.

But what stayed with me after our interview was how little joy the family finds in eating together. Food stopped meaning comfort to Karen on the day she threw up the peanut butter sandwich. And while she has worked hard to find foods she does like, it's rare that she can share them with the people closest to her. Her kids tolerate the meals she makes, but do not celebrate them; her husband almost never joins in at all. Day to day, Karen doesn't perceive this as a big loss: "We have other ways of bonding." But sometimes, it feels bittersweet to see other families cooking together, eating together, connecting over food—especially the kinds of food that are so abundant at California farmers markets, but reviled in her house. The few foods that do elicit a sense of connection in her family tend to be ones that Karen considers "unhealthy" and "processed," even though such terms make her feel bad. It is all the thick, dark smear of Nutella on toast. Peter's peculiarity has become the one food tradition that their family will pass down. So, every day, Karen tries to get dinner on the table while caught between the unattainable standards of our food culture and the reality of her family's rigid preferences. She wonders if the latter would be easier to live with if external expectations were lower. But as it is: "My relationships do not center on food," she says. "And I do not think food is love."

From 2013 to 2015, Lois Bielefeld, a photographer from Minneapolis, shot the weeknight dinners of seventy-eight families around the United States and in Luxembourg. The project took her into homes where people eat around dining room tables laden with side dishes, into apartments where single men eat at their desks, and into many, many living rooms where people eat their meals from tray tables or their laps or off the floor, while watching television. In some families, everyone was eating something different, made at a different time. In others, there were a series of takeout boxes or microwave dinners, heated up and ready to go. In many of her photos, there is joy; a sense of loved ones settling down together at the end of a hectic day. Even if they are gathering around the television, they are relaxing and together. But there is also exhaustion, tension, and a kind of shyness in most of the shots, no matter how elaborate or simple the meals.

"We have this American ideal of sitting around the table, everyone talking, everyone happy, the food is lovely," says Bielefeld. "It comes from the 1950s and 1960s, when TV started showcasing what families were doing. But it's not real family life." The "family dinner" ideal has always been a hard one to live up to, and it's now made even more complicated by the pressures of modern food culture. Bielefeld says that when she came into people's spaces, she often heard some degree of apology in how they presented themselves and in how far their reality strayed from the ideal. "People would say, 'Oh, this isn't anything special,'" she says. "But I think sharing a meal—any kind of meal—is one of the most intimate, wonderful things. And especially when it's just what they normally do."

Apologizing around food—for our failure to make it good enough, healthy enough, for what we're choosing to eat, for what we're daring to serve others—has become an important ritual in today's food culture. I heard the same kind of thing every time I shared a meal with one of the people in this book. In her beautiful DUMBO loft, Kate apologized for the mess on her kitchen counters, for her uncertainty about how long to cook a duck. In her tiny North Philly row house, Sherita offered me bottled water and apologized that it was just store brand. I also hear an apology almost every time I witness a female friend eating cheese. Or bread. Or chocolate. We feel especially compelled to apologize for enjoying food, for wanting seconds, for appearing to eat even a single bite more than we think we should.

Of course, it's always been important to show some kind of deference around food when we engage with it socially. "We have a long history of making food into a sign of civilization," says Paul

Rozin, the psychology professor at the University of Pennsylvania most known for his research on disgust and for coining the phrase "omnivore's dilemma." Rozin is fascinated by how humans have converted the animalistic act of eating into something refined and, well, palatable. He points to how most of us raised in Western society learn to use a fork and to chew with our mouths closed. "We have this way of eating where we're chewing and swallowing and yet nobody sees the food in your mouth. So you can be looking right at someone and talking through the same hole, yet manage not to show them the disgusting mess in your mouth."

But at this moment in time, most of our food apologies are rooted in diet culture and our failure to live up to its ideals. And Rozin's research shows how willing we are to make judgments about people on the basis of their diets. In one of his studies, 14 percent of women reported that they never bought chocolate in public because they worried people would judge them for such an indulgence. In another, Rozin gave survey participants a vignette citing basic facts about a fictional person: her job, her education level, and her daily diet. Half the participants were told that the imaginary woman's diet was healthy, while the other half were told the opposite. "Then we ask, 'What do you think about Jane? Is she a good person?'" Rozin tells me. "All of those opinions turn out to be related to their judgment of her diet. Food is a moral substance in America."

In some ways, this moralizing is unavoidable. "Food is such an important activity and we have so many rituals around it," says Rozin. "Think of how many people around the world give blessings or thanks at every meal. But we're also in this very complicated world, where there are many moral demands on us and we have limited time, so we have to make choices. And that's why we like simple rules, like 'Prius owners are nicer people.'" Or "People who eat vegetables are better than people who don't." And

"Women shouldn't buy—much less eat!—chocolate in public without apologizing for it."

So moralizing about food and atoning for it may be normal and time-honored traditions. But that doesn't make them good for us. These are also key ways in which we try to sever our emotional connection with food, to keep our understanding of food limited to the clinical: nutrition, health, environmental issues. And yet, it isn't our emotional connection to food—that part of our eating instinct that ties nourishment to comfort—that causes all our problems with it. It's our fear of that emotion. Think of how Nancy Zucker, the founder of the Duke Center for Eating Disorders, explained anorexia to me in Chapter 4: "There are a subset of people with eating disorders who also have a history of childhood picky eating," she noted. "But more often, I see the exact opposite: anorexic patients who loved food as kids and learned to be afraid of how much they loved food, of their emotions and the vividness of the experience. Those feelings were too powerful. It made them want to shut everything down."

So it's our discomfort—and even disgust—with the joy of eating that frightens us. And that's because of a culture that tells us, in a thousand ways, from the time we first start solid foods, that this comfort cannot be trusted. That *we* cannot be trusted to know what and how much to eat. We must outsource this judgment to experts who know better—first to our parents; then to teachers; then to food gurus and big brands, who sell us on diets, cleanses, food dogmas, and "lifestyle changes." We cede our knowledge, our own personal relationship with food, to an entire world built on the premise that we don't know how to feed ourselves.

This is how I know that it almost doesn't matter how hard we fought to preserve Violet's eating instinct, to help her rediscover it, and then, to do that again. It is nearly inevitable that she will lose that instinct over and over in the coming years. Perhaps not because of intense medical trauma this time, and (I hope) never again will she so completely lose the ability to eat that we'll need to thread another nasogastric tube down her throat. But Violet will lose touch with that innate ability to eat well for the same reasons that we all lose it. Because on play dates, our friends put their Barbie dolls on diets. Because when we wanted seconds of ice cream or birthday cake, our parents were alarmed—maybe even a little disgusted—by our ability to derive such intense pleasure from food. So now we think skinny models need to "eat a sandwich," but also that we should feel guilty for having seconds (or sometimes even firsts) of anything. Because we go to the movies and order a super-sized soda, a vat of popcorn, or the giant box of Red Vines, all designed to beguile and addict us into eating sugar, fat, and salt in quantities our brains were never designed to handle. Because we inevitably gain more weight than we want, in places we don't want it to be, and so we do a Master Cleanse or a Whole 30 challenge; we join Weight Watchers or the slow food movement. We drink protein smoothies and eat kale. We go gluten-free, vegan, dairy-free, Paleo, fat-free, macrobiotic, and we learn, over and over again, how to not eat. We are rats in a maze we built ourselves, sniffing at every new diet or food philosophy to see if this will be the path that leads us out. When really, all of those prescriptions and rules are their own kind of feeding tube, sliding into place and overriding our innate understanding of food on a slow drip. And then our mistrust of ourselves becomes a self-fulfilling prophecy. We can't trust ourselves to follow our instincts because we've never given ourselves that kind of permission to eat.

I discovered fashion magazines the year I was eleven. My best friend at summer camp was thirteen and much more worldly; we spent that summer devouring every copy of *Seventeen*, *YM*, and *Sassy* that we could get our hands on. In the back of every issue, I found the same page, one of those advertisements that are formatted like articles. Now it's easy to spot the difference, but back then, I wasn't so sure. There was no byline, but the copy was written as a first-person account of one woman's struggle with her weight—how she gained so much that her thighs rubbed together when she walked; how her boyfriend left a break-up letter on the dresser because he couldn't be with her; and how she ultimately found a "doctor-prescribed" diet that changed everything. "It's not about how much you eat, it's the particular *combinations* of food," she wrote. "Now I can eat constantly, I'm never hungry—and the weight just keeps coming off!" The piece ended with an address where you could send a check and a self-addressed stamped envelope to get your own copy of the diet plan.

I never wrote away for the plan. I was still in that happy, carefree space where I ate pretty much whatever I wanted, all the time. But I sensed that other people didn't have that freedom. My camp friend didn't talk about it, but I had heard some of our counselors whispering that she'd been treated for an eating disorder during the school year. It was the kind of mysterious detail that made her seem extra glamorous at the time. My dad weighed himself every morning and would gently chastise himself for eating any kind of dessert except on major holidays. A woman in my mom's office was obese and I often heard their co-workers gossip about why and how *that* had happened. So all around me, people were worrying about food in big and small ways. But that advertorial, tucked in the back of every teen magazine next to the horoscopes, was how I began to understand that the deepest wish of dieters everywhere was to be able to solve their weight problem while still eating whatever they wanted. And not in a sad, beleaguered "Guess the diet starts tomorrow" way, but with the easy joy that I still had around food at that age, that maybe we all remember having at some point back in childhood.

This fantasy of consequence-free eating is, of course, deeply tied to the fear that we don't know how to eat. Otherwise it wouldn't be a fantasy and we would just do it. Instead, the trope of the beautiful, skinny woman who eats like a truck driver is deeply embedded in popular culture: Lorelai Gilmore of *Gilmore Girls* fame, Grace Adler of the newly revived *Will & Grace*, and Liz Lemon on *30 Rock* all charm their fictional love interests by downing vast quantities of Chinese takeout, candy, deep-fried anything, and cheese puffs. On YouTube, a lithe Japanese woman who calls herself "the gluttonous beauty" has hundreds of thousands of people logging on to watch her binge eat, while in South Korea, thin women post videos of themselves gorging on food in an odd local trend known as *muk-bang*, which translates to "eating broadcasts." Guys love to watch a girl eat, is the message, but only when it defies the laws of physics that she could eat so much and still be so small. I often wonder whether an actress playing such a role finds it freeing. Does she really get to eat all that food on set, while shooting the scene? Or is it humiliating, to have to pretend she's so much more uninhibited than her career actually allows her to be? We may know, intellectually, that those plot lines are implausible. But that doesn't stop us from searching out diets, lifestyle plans, or food beliefs that promise that kind of freedom.

In reality, there are only a handful of people in the world who can eat absolutely anything they want and never gain weight. Abby Solomon, of Austin, Texas, is one of them. She's twenty-three years old and five feet, ten inches tall, and she weighs just 102 pounds. Because of a rare and poorly understood genetic condition called neonatal progeroid syndrome, she needs to eat all the time. "My life is kind of dominated by food," Abby says. "I constantly have granola bars in my purse. And some kind of sugar snack, in case I get hypoglycemic. It's very thought-out and methodical, which can get very annoying." Abby loves food; she enjoys cooking and eating out with her family. Tex-Mex is a favorite. But her body doesn't make enough asprosin, a hormone that regulates blood sugar. Which means she doesn't just experience hunger; if she doesn't eat every two to three hours, she's fast on the verge of going into hypoglycemic shock. The daily grind of getting enough calories in her body to prevent that takes a lot of the joy away. "When I'm eating snacks, it just feels like, 'Okay, I've got to put something in my body,'" she says.

Even though Abby eats so frequently, she can't eat large quantities of food most of the time. For reasons her doctors don't understand, her fullness signals seem to be in the same kind of overdrive as her hunger cues, so she starts to feel stuffed after half a hamburger or a slice of French toast. "Some days I can eat more and feel okay. The problem is, even when I do eat a lot, I still get hungry an hour later because my metabolism is so fast," she notes. She eats cheese and crackers before bed, because otherwise she'll wake up shaky with hunger in the middle of the night. But it's a constant dance: Eat enough to take away the sweaty, jittery feeling of starvation, but not so much that she becomes nauseated. "Sometimes I do skip a snack or stop eating before I'm really full because I just don't want to eat anymore," she says. "I feel weird saying that because it's like, 'Why do you stop eating if you're not full?' But I don't know. I'll just have to do this again in an hour. Maybe finishing this will buy me an extra thirty minutes. But there's really no difference."

The last years of elementary school and pretty much all of middle school were rough for Abby. Kids began noticing that she got to have snacks at times when nobody else was allowed—

in class, in her bunk at sleep-away camp—and even more, they noticed her appearance. "I looked more severe then," Abby says. "I'm still bony, but as a kid I was really thin and bony, and also tall." It was not easy to blend in. She would have to explain that she had a rare syndrome, but beyond that, she didn't really know how to respond to other children's questions or their teasing. "There were definitely curious stares and mean stares and I'm sure I was bullied," she says. "I've kind of blocked it all out."

At times, the other consequences of Abby's syndrome trump any concerns about her weight and relationship with food. In 2015, she had to drop out of college because her vision was deteriorating. She underwent four surgeries to have her nasal passages rebuilt and the lenses in each eye replaced to prevent blindness, and still deals with chronic eye issues. This is one price, then, of being able to eat anything. The other is that the world around you will never let you forget how much you deviate from the norm. "I was getting a pedicure and the nail tech saw my thin ankles and was like, 'Get some meat on those bones!'" she tells me. "It was kind of sweet because she offered me her leftover ham from Christmas. But I dislike that expression. I do like that I can eat and not gain weight, but there's a pro and a con to everything. It's hard to have people constantly wondering whether I'm anorexic or telling me I need to eat more."

There have been times when Abby was out with friends and became aware of a subtle tension. "With a bunch of girlfriends, you know, someone is going to say, 'I want the bread, but oh no, I can't have it,'" she explains. "And they'll say they're so jealous of me." She's worked hard to get past feeling self-conscious about ordering the burger when others are getting salad. And Abby "hasn't dated anyone, like, ever" and admits to worrying what a potential romantic partner will think about her appearance and food choices. "For women, there's that whole thing of, 'don't seem too hungry, don't eat too much, don't be gross.'" Even when you live the fantasy, it seems, you might still internalize the fear.

Teaching Violet to eat again was like reintroducing an animal to the wild. We were asking her to eat when we weren't sure how to do it ourselves; to listen to an instinct that neither Dan nor I could always reliably hear in our own bodies. There were times when one or all of us wanted to run back to the confines and comfort of the tube, the doctor-ordered diet, the idea that someone else knows better. That same fear—that we can't trust ourselves around food— is what drives so many of us when we stop eating sugar or start yet another diet. We'd rather subscribe to eating as an endless cycle of punishment and reward than struggle to decide for ourselves what to eat, and when, and how much. And we give up that free-dom of choice without letting go of our guilt. Someone else decides what it means to "eat right," decides that we must eat our vegeta-bles before we can have dessert. But when we can't do it, the fail-ure is all ours. This fixation on willpower is at the root of all weight stigma: large bodies are a constant reminder of our fear of this weakness, this lack of control.

At the same time, I am encouraged by what seems to be a nascent but growing awareness in certain corners that weight is not about willpower and that diets fail us, not the other way around. As I write this chapter, *The New York Times Magazine* has just pub-lished a piece about the tumbling revenues of Weight Watchers and the fifty-four-year-old diet behemoth's efforts to rebrand itself as a "lifestyle program" that offers benefits "beyond the scale." Of course, there's no getting around that name; why watch your weight if you aren't hoping to see it change? But dieting is now "consid-ered tacky," writes the journalist Taffy Brodesser-Akner, of the cultural shift fueling the company's makeover. "'Weight loss' was a pursuit that had, somehow, landed on the wrong side of political correctness. People wanted nothing to do with it. Except that many of them did: They wanted to be thinner. They wanted to be not quite so fat. Not that there was anything wrong with being fat! They just wanted to call dieting something else entirely." Yet this very paradox—that we'll accept our bodies, as long as they're not too fat—underscores how much we still subscribe to the same thin ideal, the same belief that we can't get there on our own. We're still looking for the plan, only now the plan is organic, artisanal, more expensive, more steeped in socioeconomic privilege. The backlash against dieting has, so far, only raised the bar on how to "eat right."

Maybe this is all part of a messy transition. We are starting to recognize that the rise of modern diet culture has helped to create the very obesity epidemic it purports to solve. We aren't surprised when we hear statistics like the fact that over the course of five years, two thirds of dieters will regain more weight than they initially lost. That finding comes from an evidence review by UCLA psychologists; their research also shows that dieters gain more, on average, in a two-year period than people not trying to lose weight. And that news doesn't shock us either. We're also starting, perhaps more tentatively, to question the infallibility of the alternative-food movement, to see the flaws in requiring $8 heirloom tomatoes and hours of intensive home cooking as essential to any solutions for our broken food system. We're finally challenging the twin ideologies around eating that have so defined our food culture for the past thirty years. Maybe the next step will be to start to listen to ourselves, to realize that these broader cultural shifts can apply to what's in our very own refrigerator, to how we feel about what we're eating for lunch today.

This will not be a quick or easy process. To the psychologists, doctors, nutrition scientists, and advocates who champion the Health at Every Size approach, just as to the disciples of Satter's Division of Responsibility, the answer seems simple: Eat the type and amount of food you want, when you want it. Recognize that all bodies are valuable and worthy of respect. Decide you can make choices for your health without making a moral judgment about your weight. View the goals of nutrition and a more sustainable food system as worthwhile, but not so all-encompassing that they should dictate how you behave at every meal.

But so many parts of this new approach are unnerving. Can we really relinquish the belief that there is a moral difference between eating vegetables from the farmers market and Tastykakes from the corner store? For that matter, how do we even know which one we really want, as opposed to the one that we've been programmed to like? We have to get reacquainted with our own innate prefer-ences. We must decide for ourselves what we like and dislike, and how different foods make us feel when we aren't prejudging every bite we take. It takes its own kind of relentless vigilance to screen out all that noise. It requires accepting that the weight you most want to be may not be compatible with this kind of more intuitive eating—but that it's nevertheless okay to be this size, to take up the space that your body requires.

All of that is hard. Still, I do believe that it's possible for anyone—infant, child, teenager, adult—to sense their own hunger and fullness, and to eat on their own terms, for both pleasure and health. And, in doing so, to move toward valuing their bodies for reasons beyond the aesthetic. Helping Violet learn to eat again brought me closer to being able to do it than ever before in my life. During the months when she was first embracing food—and first learning to talk and to listen—I became acutely aware of how often adults denigrate food and their own bodies, in front of children. "Oh, Violet, you're so lucky you can eat cheese!" a well-meaning caregiver said to my fifteen-month-old one afternoon as they pushed cubes of cheddar around on her high chair tray. "I love cheese, but it makes me fat." I cringed, but some weeks later, I was the one apologizing, when Dan gave Violet a piece of bacon to chew on during a session with Maggie Ruzzi, her dietitian. "Great, now Maggie will think we let our kid eat bacon all the time!" I joked. Ruzzi stopped me: "Violet, bacon is one of the foods that makes life worth living," she said. Another day, I expounded on some sartorial frustration at the dinner table—the jeans that didn't fit or the pregnancy weight that never quite vanished—and said, "I'm just not that happy with my body right now." In response, and most likely just because she recognized a few words, Violet began patting herself all over, saying "My body! My body!" That's when I heard it—and every other time I've ranked foods, shamed myself for gravitating toward the "wrong" foods, or criticized my body out loud. And so I stopped.

There are, of course, times when I feel compelled to explain to Violet why we won't be having ice cream with dinner after a day that featured, say, breakfast doughnuts and then cupcakes at a friend's birthday party. But I try to frame it in ways that don't demonize those foods. "Treats are great, but we need to eat lots of different kinds of foods to feel strong," I'll offer. "We wouldn't feel good if we ate broccoli at every meal, either." What I want her to hear: "Ice cream and cupcakes are not bad foods. There are no bad foods." And still, I wrestle with that subtle distinction. Ice cream is not broccoli. That is an indisputable fact. And it is also an emotional and moral distinction made by our food culture, for reasons that are far from fact-based.

There have been other changes too. Some, like having regular family dinners, probably would have happened regardless, as Dan and I transitioned from our less structured pre-kids existence (when dinner was most often eaten late and in front of Netflix) to the more exacting rhythms of family life. We eat breakfast together most mornings too—a meal that Dan, in particular, used to consume while racing to work or not at all. This was a conscious shift made on Violet's behalf, so she could see us modeling mealtimes more than once a day. But we've learned that it isn't just Violet who does better on a regular meal schedule. We're also less cranky, more functional, and more able to suss out our own hunger and fullness cues than when we skip meals or graze all day.

Just as I begin immersing myself in researching and writing the early chapters of this book, I get pregnant again. Eating during this second pregnancy is very different from eating during the first. I know now that everything I put myself through the first time—the odd-tasting herbal shakes prescribed by my acupuncturist, the relentless counting of protein grams and calcium servings, the religious abstention from caffeine and nitrates—wasn't enough to ensure a healthy baby. We have only moderate control over the outcome of any pregnancy. And so there is far less obsessing over perfect nutrition, far less thinking of food as the building blocks for a perfect baby. And there is much more eating according to my actual hunger cues and cravings, even on days when that means I eat nothing but toast and plain pasta. The height of my morning sickness coincides with reporting Chapter 4 and seems to serve that research well. Although I'm normally an adventurous eater, stymied by the rigidity of the adult picky eaters in my own life, I feel a new and profound empathy when Marisa, Ben, Jennifer, and others describe the revulsion they experience when they're trying to force themselves to eat beyond their list of safe foods. I'm having the same knee-jerk response anytime I see a package of raw meat or contemplate eating a salad.

Then it is a summer Saturday afternoon, late in my third trimester. We're reading, snuggled in bed together before her afternoon nap, when Violet asks me, rather offhandedly, how her baby sister will eat. "I'll probably feed her milk with my breast," I explain, trying desperately to sound casual. The thought of trying to breast-feed again has haunted me for months now. I want it to work so badly; to find some kind of healing in the knowledge that I can do this thing without accidentally starving a baby, without missing all the signs of her struggle to breathe, let alone eat. And yet even more, I need contingency plans upon contingency plans, so this time it doesn't feel like this feeding business rests exclusively on my shoulders. There are already two containers of newborn formula secreted in my bag for the hospital. I have ordered a new breast pump, but can't bring myself to open the box. "And you and Daddy can feed her another kind of milk with bottles. So she will eat milk, from my breast or from a bottle."

Violet thinks about that. "Or I can feed her with my feeding tube!" she suggests.

I pause. We're never sure what Violet remembers about her feeding-tube experiences. She has a little doctor's kit in her playroom, stocked with a mix of pretend Fisher-Price medical supplies and the real deal: a feeding tube, chest drain, blood pressure cuff, and oxygen mask. We've included the latter because everything you read about helping a child navigate frequent medical interventions emphasizes the importance of play therapy, of letting them act out whatever memories or questions are on their minds. Violet often plays doctor with an accuracy few four-year-olds can supply. She takes our temperatures, gives shots, and kisses boo-boos, yes. But she also frequently intubates her teddy bear "because Oso needs help breathing." She suctions blood out of her baby doll's mouth. And she uses the kit's feeding tube to nourish her toys with imaginary mac and cheese and ice cream.

"I don't think your baby sister will use a feeding tube," I say carefully. "Most babies don't eat that way."

Violet ponders this. She is unsatisfied. "But maybe she can have my old feeding tube?" she asks. "I can share it with her. And then, when she goes to the hospital when she's two, maybe the doctors will give her one all for her own."

I cannot answer. This is one of what I am sure will be at least one million moments when I realize my daughter is smarter than I am. To me, the feeding tube has symbolized so much shame and sadness. It was what marked her as a more fragile, "less able" child and marked me as a mother who had failed. But it is none of those things to Violet. It's just the way she ate as a baby. The tube is not good or bad to her; it doesn't rank below the bottle or the breast. It left a scar on her abdomen, but somehow, not on her sense of who she is. It is something she knows, a part of her story. And as a four-year-old who knows all about the importance of sharing and fairness, she worries that her baby sister will be somehow deprived if she does not have the chance to know it too.

Violet's thinking on this front may change as she grows up, just as her ability to tell the difference between physical hunger and "I see a cookie" hunger is already clouding over. But what if we could think about all kinds of eating as clearly as she thinks about the tube right now? Without judgment. Without guilt. Without ranking picky eaters as somehow less than adventurous eaters, corner stores as less than farmers markets, meat eaters as less than vegetarians, fat as less than thin. What would our food culture look like then? Eating disorders would still exist, of course. Food is too often only the symptom of a larger problem—the collateral damage of open-heart surgery or genetic predispositions. Our biology would still program our bodies with set points and we'd still have to reckon with how that creates a diversity of human body types that don't all measure up to what our culture defines as beautiful. And inequality would still exist; children would still grow up in households with empty kitchen cupboards. But maybe the process of eating, at least, would be less fraught. Maybe we'd be less inclined to classify certain foods as addictive and to punish ourselves for loving them. We would seek fewer false idols of nutrition and wellness. We wouldn't need them, if we believed that everyone has their own innate understanding of how to eat.

Violet didn't know how to eat until we took away the tube. But only four years later do I realize that it was really me who couldn't recognize how much she did know even while she was learning to master chewing, swallowing, and other mundane aspects of the job. It took Violet's feeding tube for me to understand how much we've all trained ourselves to ignore our own instincts, to be uncomfortable with our own hunger and the pleasures of food. But what is lost can be found. And then, lost and found again. Recognizing ourselves as capable eaters means identifying the factors that caused us to lose that identity in the first place—the particular mix of biology, psychology, socioeconomic positioning, and life experience that is different for everyone. It means reclaiming control of our bodies. And it means accepting that this is an ongoing process, one we'll begin again at every meal.

The only way to learn to eat is by eating.

NOTES AND SOURCES

This book is the result of my interviews with researchers, scientists, doctors, dietitians, advocates, and all kinds of people who feed themselves and their families every day. In addition to talking to people directly, I read dozens of studies, articles, books, and other published materials about food, which provided the statistics and research details referenced in these pages, or more generally informed my thinking about the issues. I also conducted online surveys and spent hours reading threads in Facebook groups and other internet forums devoted to the various food issues and subcultures explored here. And I ate a lot of meals.

The majority of this reporting was done specifically for this book. I also drew on previous conversations with sources interviewed for past articles on related topics. Those articles are cited chapter by chapter below. The details of Violet's story are drawn from conversations with her doctors and therapists, medical records, and my own obsessive note-taking and journaling during her various hospital stays and therapy programs. It should be acknowledged that notes kept during those times were not fact-checked and I was often writing about rapidly unfolding events while in the midst of shock and trauma. I've done my best to cross-reference my own records and memories with Dan and others close to our situation, as well as with the medical professionals who have provided Violet with such meticulous care. Any errors are entirely my own.

CHAPTER ONE

The foundation for this chapter was my story "When Your Baby Won't Eat," published in The New York Times Magazine *on February 4, 2016.*

STATISTICS ON THE FREQUENCY OF PEDIATRIC FEEDING PROBLEMS:

Data collected by the Feeding Tube Awareness Foundation (feedingtubeawareness.org) and Feeding Matters (feedingmatters.org).

Reau, Nancy, Yvonne D. Senturia, Susan A. Lebailly, and Katherine Kaufer Christoffel. "Infant and Toddler Feeding Patterns and Problems: Normative Data and a New Direction." *Journal of Developmental & Behavioral Pediatrics* 17 3 (June 1996): 149–53.

ANALYSIS OF HOW A CHILD'S EATING INSTINCTS ARE DISTORTED:

Canetti, Laura, Eytan Bachar, and Elliot M. Berry. "Food and Emotion." *Behavioural Processes* 60 2 (2002): 157–64.

Davis, Clara. "Self Selection of Diet by Newly Weaned Infants: An Experimental Study." *American Journal of Diseases of Children* 36 4 (1928): 651–79.

Fiese, Barbara H., and Marlene Schwartz. "Reclaiming the Family Table: Mealtimes and Child Health and Wellbeing." *Social Policy Report* 22 4 (2008): 1–19.

Frankel, Leslie Ann, Sheryl O. Hughes, Teresia M. O'Connor, Thomas G. Power, Jennifer Orlet Fisher, and Nancy L. Hazen. "Parental Influences on Children's Self-Regulation of Energy Intake: Insights from Developmental Literature on Emotion Regulation." *Journal of Obesity* (2012).

Galloway, Amy T., Laura M Fiorito, Lori Francis, and Leann Lipps Birch. "'Finish Your Soup': Counterproductive Effects of Pressuring Children to Eat on Intake and Affect." *Appetite* 46 3 (2006): 318–23.

Ong, K. K., M. L. Ahmed, P. M. Emmett, M. A. Preece, and D. B. Dunger. "Association Between Postnatal Catch-up Growth and Obesity in Childhood: Prospective Cohort Study." *BMJ* 320 7240 (2000): 967–71.

Savage, Jennifer S., Jennifer Orlet Fisher, and Leann Lipps Birch. "Parental Influence on Eating Behavior: Conception to Adolescence." *Journal of Law, Medicine & Ethics* 35 1 (2007): 22–34.

RESEARCH ON BEHAVIORAL MODEL OF PEDIATRIC FEEDING THERAPY:

Müller, Michael, Cathleen C. Piazza, James W. Moore, Michael E. Kelley, Stephanie A. Bethke, Angela E. Pruett, Amanda J. Oberdorff, and Stacy A. Layer. "Training Parents to Implement Pediatric Feeding Protocols." *Journal of Applied Behavior Analysis* 36 4 (2003): 545–62.

Sharp, William G., David L. Jaquess, Jane F. Morton, and Caitlin V. Herzinger. "Pediatric Feeding Disorders: A Quantitative Synthesis of Treatment Outcomes." *Clinical Child and Family Psychology Review* 13 4 (2010): 348–65.

Strologo, Luca dello, F. Principato, Donatella Sinibaldi, Aldo Claris Appiani, Fabiola Terzi, Anne Marie Dartois, and Gianfranco Rizzoni. "Feeding Dysfunction in Infants with Severe Chronic Renal Failure after Long-Term Nasogastric Tubefeeding." *Pediatric Nephrology* 11 (1997): 84–86.

Volkert, Valerie M., Kathryn M. Peterson, Jason R. Zeleny, and Cathleen C. Piazza. "A Clinical Protocol to Increase Chewing and Assess Mastication in Children with Feeding Disorders." *Behavior Modification* 38 5 (2014): 705–29.

Wilkins, Jonathan W., Cathleen C. Piazza, Rebecca A. Groff, and Petula C. M. Vaz. "Chin Prompt Plus Re-Presentation as Treatment for Expulsion in Children with Feeding Disorders." *Journal of Applied Behavior Analysis* 44 3 (2011): 513–22.

40 PERCENT OF TEENAGE GIRLS ARE USING "RESTRICTIVE MEASURES" TO LOSE WEIGHT:

Micali, Norberto Liborio, Bianca Lucia De Stavola, George Basil Ploubidis, Emily Simonoff, Jennifer D. Treasure, and Alison E. Field. "Adolescent Eating Disorder Behaviours and Cognitions: Gender-Specific Effects of Child, Maternal and Family Risk Factors." *British Journal of Psychiatry* (2015).

36.5 PERCENT OF AMERICANS STRUGGLE WITH OBESITY:

Ogden, Cynthia L., Margaret D. Carroll, Cheryl D. Fryar, and Katherine M. Flegal. "Prevalence of Obesity Among Adults and Youth: United States, 2011–2014." *National Center for Health Statistics Data Brief* 219 (2015): 1–8.

RESEARCH ON THE CHILD-LED MODEL OF FEEDING THERAPY:

Chatoor, Irene. *Diagnosis and Treatment of Feeding Disorders in Infants, Toddlers and Young Children.* Washington, D.C.: National Center for Clinical Infant Programs, 2009.

Fox, Mary Kay, Barbara L. Devaney, Kathleen C. Reidy, Carol M. Razafindrakoto, and Paula J. Ziegler. "Relationship Between Portion Size and Energy Intake Among Infants and Toddlers: Evidence of Self-Regulation." *Journal of the American Dietetic Association* 106 1 Suppl. 1 (2006): 77–83.

Morris, Suzanne, and Marsha Dunn Klein. *Pre-Feeding Skills: A Comprehensive Resource for Mealtime Development.* Austin, TX: Pro Ed, 2000.

Rowell, Katja, and Jenny McGlothlin. *Helping Your Child with Extreme Picky Eating: A Step-by-Step Guide for Overcoming Selective Eating, Food Aversion, and Feeding Disorders.* Oakland, CA: New Harbinger Publications, 2015.

Satter, Ellyn. *Child of Mine: Feeding with Love and Good Sense*. Boulder, CO: Bull Publishing, 2000.

———. "The Feeding Relationship: Problems and Interventions." *Journal of Pediatrics* 117 2 Pt. 2 (1990): 181–89.

———. *How to Get Your Kid to Eat . . . But Not Too Much*. Boulder, CO: Bull Publishing, 1987.

Tribole, Evelyn, and Elyse Resch. *Intuitive Eating: A Revolutionary Program That Works*, 3rd ed. New York: St. Martin's Griffin, 2012.

INFORMATION ON BLENDED TUBE-FEEDING DIETS:

Klein, Marsha Dunn, and Suzanne Evans Morris. *Homemade Blended Formula Handbook*. Tucson, AZ: Mealtime Notions, 2007.

BREAST-FEEDING VERSUS FORMULA:

Division of Nutrition and Physical Activity: Research to Practice Series No. 4: *Does Breastfeeding Reduce the Risk of Pediatric Overweight?* Atlanta: Centers for Disease Control and Prevention, 2007.

Evenhouse, Eirik, and Siobhan C. Reilly. "Improved Estimates of the Benefits of Breastfeeding Using Sibling Comparisons to Reduce Selection Bias." *Health Services Research* 40 6 Pt. 1 (2005): 1781–802.

Hediger, Mary L., Mary D. Overpeck, Robert J. Kuczmarski, and W. June Ruan. "Association Between Infant Breastfeeding and Overweight in Young Children." *JAMA* 285 19 (2001): 2453–60.

Kramer, Michael S., Lidia Matush, Irina Vanilovich, Robert W. Platt, Natalia Bogdanovich, Zinaida Sevkovskaya, Irina Dzikovich, Gyorgy Shishko, Jean-Paul Collet, Richard M. Martin, George Davey Smith, Matthew W. Gillman, B. L. Chalmers, Ellen D. Hodnett, and Stanley Shapiro. "A Randomized Breast-Feeding Promotion Intervention Did Not Reduce Child Obesity in Belarus." *Journal of Nutrition* 139 2 (2009): 417S–21S.

Jung, Courtney. *Lactivism: How Feminists and Fundamentalists, Hippies and Yuppies, and Physicians and Politicians Made Breastfeeding Big Business and Bad Policy*. New York: Basic Books, 2015.

Quart, Alissa. "The Milk Wars." *New York Times*, July 14, 2012.

Rosin, Hanna. "The Case Against Breastfeeding." *The Atlantic*, April 2009.

DEVELOPMENT OF SWEET FLAVOR PREFERENCES IN CHILDHOOD:

Drewnowski, A., Julie A. Mennella, Susan L. Johnson, and France Bellisle. "Sweetness and Food Preference." *Journal of Nutrition* 142 6 (2012): 1142S–8S.

Liem, Djin Gie, and C. N. de Graaf. "Sweet and Sour Preferences in Young Children and Adults: Role of Repeated Exposure." *Physiology & Behavior* 83 3 (2004): 421–9.

CHAPTER TWO

THIS CHAPTER WAS INFORMED BY SOME OF MY PREVIOUS WRITING:

Sole-Smith, Virginia. "Can You Really Detox?" *SELF*, July 2015.

————. "The Dairy Debate." *SELF*, October 2015.

————. "Hey, Hey, Hey Sugar!" *Women's Health*, April 2007.

THE FOLLOWING SERVED AS GENERAL SUBJECT MATTER RESEARCH FOR THIS CHAPTER:

Bratman, Steven. "What Is Orthorexia?" *Orthorexia*, January 23, 2014. http://www.orthorexia.com/what-is-orthorexia/.

Harrington, Christy. *Food Psych Podcast*. https://christyharrison.com /foodpsych/.

Oyston, Glenys. "Food Is Not Medicine." *Dare to Not Diet* (blog), April 18, 2016. https://daretonotdiet.wordpress.com/2016/04/18/food-is -not-medicine/.

Pollan, Michael. *The Omnivore's Dilemma: A Natural History of Four Meals*. New York: Penguin, 2007.

Reba-Harrelson, Lauren, Ann F. Von Holle, Robert M. Hamer, R. Mark Swann, Manuel Reyes, and C. M. Bulik. "Patterns and Prevalence of Disordered Eating and Weight Control Behaviors in Women Ages 25–45." *Eating and Weight Disorders* 14 4 (2009): e190–98.

Schlosser, Eric. *Fast Food Nation: The Dark Side of the All-American Meal*, rev. ed. Boston: Mariner Books, 2012.

Spurlock, Morgan. *Super Size Me*. 2004. Culver City, CA: Columbia Tri-Star Home Entertainment.

Tandoh, Ruby. "The Unhealthy Truth Behind 'Wellness' and 'Clean Eating.'" *VICE*, May 13, 2016.

Waters, Alice (@AliceWaters). "A message for @JeffBezos!" Twitter, June 16, 2017, 4:52 p.m. https://twitter.com/AliceWaters/status/875863664 170909696.

DATA ON FOOD INTOLERANCES:

National Institutes of Diabetes and Digestive and Kidney Diseases. *Definitions & Facts for Celiac's Disease*, June 2016. https://www.niddk.nih.gov /health-information/digestive-diseases/celiac-disease/definition-facts/.

National Institutes of Health, U.S. National Library of Medicine, Genetics Home Reference. *Lactose Intolerance*. https://ghr.nlm.nih.gov/condition /lactose-intolerance/.

DATA ON THE SIZE OF THE ALTERNATIVE-FOOD AND WELLNESS INDUSTRY:

Belluz, Julia. "The Most Googled Diet in Every City." *Vox*, November 10, 2015.

Chen, Angus. "Diet Foods Are Tanking. So the Diet Industry Is Now Selling 'Health.'" NPR, January 20, 2016.

Global Wellness Institute. "Statistics and Facts." https://www.globalwellness institute.org/press-room/statistics-and-facts/.

LaRocca, Amy. "The Wellness Epidemic." *The Cut*, June 27, 2017.

Packaged Facts. *Meal Kit Delivery Services in the U.S.*, April 28, 2016. https://www.packagedfacts.com/Meal-Kit-Delivery-10037319/.

Shepardson, David, and Jeffrey Dastin. "Amazon Deal for Whole Foods Wins US Regulatory, Shareholder Approvals." Reuters, August 23, 2017.

RESEARCH ON FOOD SENSITIVITIES AND ELIMINATION DIETS:

The LEAP Diet is marketed by Oxford Biomedical Technologies at NowLEAP.com.

Carr, Stuart, Edmond Chan, Elana Lavine, and William Moote. "CSACI Position Statement on the Testing of Food-Specific IgG." *Allergy, Asthma & Clinical Immunology* 8 1 (2012): 12.

Centers for Disease Control and Prevention. *Fourth National Report on Human Exposure to Environmental Chemicals: Executive Summary*, 2009.

———. *Fourth National Report on Human Exposure to Environmental Chemicals: Updated Tables*, 2017.

Gershon, Michael. *Your Second Brain: A Groundbreaking New Understanding of Nervous Disorders of the Stomach and Intestine*. New York: Harper Perennial, 1999.

Hecht, Susanna, F. L. Chung, John P. Richie, Shobha A. Akerkar, A. Borukhova, Lukasz Skowronski, and Steven G. Carmella. "Effects of Watercress Consumption on Metabolism of a Tobacco-Specific Lung Carcinogen in Smokers." *Cancer Epidemiology, Biomarkers & Prevention* 4 8 (1995): 877–84.

Klein, Andreas, and Hosen Kiat. "Detox Diets for Toxin Elimination and Weight Management: A Critical Review of the Evidence." *Journal of Human Nutrition and Dietetics* 28 6 (2015): 675–86.

Merrell, Woodson. *The Detox Prescription: Supercharge Your Health, Strip Away Pounds and Eliminate the Toxins Within*. New York: Rodale, 2013.

Shilton, A. C. "Should You Take a Food Sensitivity Test?" *Outside*, November 17, 2016.

INFORMATION ABOUT MULTIPLE SCLEROSIS AND DIET:

Roy Swank story via the Swank MS Foundation. www.swankmsdiet.org.

McKelvey, Cynthia. "Does Diet Matter in Multiple Sclerosis?" *Multiple Sclerosis Discovery Forum*, November 25, 2014. http://www.msdiscovery .org/news/news_synthesis/15345-does-diet-matter-multiple-sclerosis/.

Wahls, Terry. *The Wahls Protocol: A Radical New Way to Treat All Chronic Immune Conditions Using Paleo Principles.* New York: Avery, 2014. (See Terry Wahls.com for original subtitle.)

RESEARCH ON DISORDERED EATING PATTERNS AMONG EATING-DISORDER AND WELLNESS PROFESSIONALS:

Bratman, Steven. *Health Food Junkies: Orthorexia Nervosa, Overcoming the Obsession with Healthy Eating.* New York: Harmony, 2004.

Gianini, Loren M., B. Timothy Walsh, Joanna E. Steinglass, and Laurel Mayer. "Long-Term Weight Loss Maintenance in Obesity: Possible Insights from Anorexia Nervosa?" *International Journal of Eating Disorders* 50 4 (2017): 341–342.

Kassier, Suna Maria, and F. J. Veldman. "Eating Behaviour, Eating Attitude and Body Mass Index of Dietetic Students Versus Non-Dietetic Majors: A South African perspective." *South African Journal of Clinical Nutrition* 27 3 (2014): 109–13.

Ozeneoglu, Aliye, Gökce Unal1, Aydan Ercan, Hatice Kumcagiz, and Kamil Alakus. "Are Nutrition and Dietetics Students More Prone to Eating Disorders Related Attitudes and Comorbid Depression and Anxiety Than Non-Dietetics Students?" *Food and Nutrition Sciences* 6 (2015): 1258–66.

Puhl, Rebecca M., Joerg Luedicke, and Carlos M. Grilo. "Obesity Bias in Training: Attitudes, Beliefs and Observations Among Advanced Trainees in Professional Health Disciplines." *Obesity* 22 4 (2014): 1008–15.

Puhl, Rebecca M., Janet D. Latner, Kelly M. King, and Joerg Luedicke. "Weight Bias Among Professionals Who Treat Eating Disorders: Associations with Attitudes About Treatment and Perceptions of Patient Outcomes." *International Journal of Eating Disorders* 47 1 (2014): 65–75.

Puhl, Rebecca M., Dianne R. Neumark-Sztainer, Steve Austin, Joerg Luedicke, and Kelly M. King. "Setting Policy Priorities to Address Eating Disorders and Weight Stigma: Views from the Field of Eating Disorders and the US General Public." *BMC Public Health* 14 (2014): 524–33.

Yu, Zhiping, and Michael Tan. "Disordered Eating Behaviors and Food Addiction Among Nutrition College Students." *Nutrients* 8 11 (2016): 673.

RESEARCH ON RELATIONSHIP BETWEEN DIGESTIVE DISORDERS AND EATING DISORDERS:

Abraham, Suzanne, and John E. Kellow. "Do the Digestive Tract Symptoms in Eating Disorder Patients Represent Functional Gastrointestinal Disorders?" *BMC Gastroenterology* 13 (2013): 38.

Evans, Marci. "Digestive Disorders and Eating Disorders: A Complicated Mix." *Marci RD Nutrition* (blog), March 6, 2017. https://marcird.com /blog/digestive-disorders-eating-disorders-a-complicated-mix/.

Janssen, Pieter H. H. "Can Eating Disorders Cause Functional Gastrointestinal Disorders?" *Neurogastroenterology and Motility* 22 12 (2010): 1267–69.

CHAPTER THREE

THE FOLLOWING SERVED AS GENERAL SUBJECT MATTER RESEARCH FOR THIS CHAPTER:

Castle, Jill, and Maryann Jacobson. *Fearless Feeding.* Hoboken, NJ: Wiley, 2013.

Wilson, Bee. *First Bite: How We Learn to Eat.* New York: Basic Books, 2016.

STATISTICS ON BREAST-FEEDING FREQUENCY:

Lansinoh. *The 2015 Lansinoh Global Breastfeeding Survey,* July 27, 2015.

RESEARCH ON THE BUSINESS OF BREASTFEEDING, PRENATAL AND POSTNATAL NUTRITION:

La Leche League International. *2015–2016 Annual Report.*

STATISTICS ON FEMALE BODY AND WEIGHT DISSATISFACTION:

Coker, Elise L., and Suzanne S. Abraham. "Body Weight Dissatisfaction Before, During and After Pregnancy: A Comparison of Women With and Without Eating Disorders." *Eating & Weight Disorders* 20 1 (2015) 71–9.

Frederick, David A., Gaganjyot Sandhu, Patrick Morse, and Viren Swami. "Correlates of Appearance and Weight Satisfaction in a U.S. National Sample: Personality, Attachment Style, Television Viewing, Self-Esteem, and Life Satisfaction." *Body Image* 17 (2016): 191–203.

RESEARCH ON THE ROLE OF WEIGHT IN PRENATAL, POSTPARTUM, AND INFANT HEALTH:

American College of Obstetricians and Gynecologists. *Committee Opinion: Weight Gain During Pregnancy*, January 2013.

Johnsson, I. W., Birgitta Haglund, Fredrik S. E. Ahlsson, and J. Gustafsson. "A High Birth Weight Is Associated with Increased Risk of Type 2 Diabetes and Obesity." *Pediatric Obesity* 10 2 (2015): 77–83.

Lord, Catherine. "Fetal and Sociocultural Environments and Autism." *American Journal of Psychiatry* 170 4 (2013): 355 58.

Wilcox, Alex. "On the Importance—and the Unimportance—of Birthweight." *International Journal of Epidemiology* 30 6 (2001): 1233–41.

RESEARCH ON INFANT FLAVOR LEARNING AND CHILDHOOD PICKY EATING:

Aigueperse, Nadège, Ludovic Calandreau, and Aline Bertin. "Maternal Diet Influences Offspring Feeding Behavior and Fearfulness in the Precocial Chicken." *PLOS ONE* (2013).

Birch, Leann Lipps. "Development of Food Acceptance Patterns in the First Years of Life." *Proceedings of the Nutrition Society* 57 4 (1998): 617–24.

Carruth, Betty Ruth, Paula J. Ziegler, Anne K. Gordon, and S. Barr. "Prevalence of Picky Eaters Among Infants and Toddlers and Their Caregivers'

Decisions About Offering a New Food." *Journal of the American Dietetic Association* 104 1 Suppl. 1 (2004): s57–64.

Lafraire, Jérémie, Camille Rioux, Agnès Giboreau, and Delphine Picard. "Food Rejections in Children: Cognitive and Social/Environmental Factors Involved in Food Neophobia and Picky/Fussy Eating Behavior." *Appetite* 96 (2016): 347–57.

Lewinsohn, Peter M., Jill M. Holm-Denoma, Jeffrey M. Gau, Thomas E. Joiner, Ruth H. Striegel-Moore, Patty Bear, and Becky Lamoureux. "Problematic Eating and Feeding Behaviors of 36-Month-Old Children." *International Journal of Eating Disorders* 38 3 (2005): 208–19.

Mascola, Anthony Joseph, Susan W. Bryson, and William Stewart Agras. "Picky Eating During Childhood: A Longitudinal Study to Age 11 Years." *Eating Behaviors* 11 4 (2010): 253–57.

Mennella, Julie. "The Chemical Senses and the Development of Flavor Preferences in Humans;" in P. E. Hartmann and Hale T. (eds). *Textbook on Human Lactation*. Plano, TX: Hale Publishing, 2007, 403–14.

Rollins, Brandi Y., Eric Loken, Jennifer S. Savage, and Leann Lipps Birch. "Effects of Restriction on Children's Intake Differ by Child Temperament, Food Reinforcement, and Parent's Chronic Use of Restriction." *Appetite* 73 (2014): 31–9.

Skye Van Zetten's blog is mealtimehostage.com. Her Facebook group is also called Mealtime Hostage and can be found at facebook.com/groups /MealtimeHostage/.

THE BENEFITS OF MAINSTREAMING SOCIALLY CONSCIOUS, ENVIRONMENTALLY FRIENDLY HEALTH FOOD:

U.S. Department of Agriculture, Agricultural Marketing Service. *Supporting Local & Regional Food Systems: Helping American Farmers Feed the Country*, April 2016.

CHAPTER FOUR

THE FOLLOWING SERVED AS GENERAL SUBJECT MATTER RESEARCH FOR THIS
CHAPTER:

Kohn, Jill Balla. "What Is ARFID?" *Journal of the Academy of Nutrition and Dietetics* 116 11 (2016): 1872.

Batsell, W. Robert, A. Shane Brown, Matthew E. Ansfield, and Gayla Y. Paschall. "'You Will Eat All of That!': A Retrospective Analysis of Forced Consumption Episodes." *Appetite* 38 3 (2002): 211–19.

Birch, Leann Lipps, Jennifer Orlet Fisher, and Kirsten Krahnstoever Davison. "Learning to Overeat: Maternal Use of Restrictive Feeding Practices Promotes Girls' Eating in the Absence of Hunger." *American Journal of Clinical Nutrition* 78 2 (2003): 215–20.

Galloway, Amy T., Laura M. Fiorito, Yoonna Lee, and Leann Lipps Birch. "Parental Pressure, Dietary Patterns and Weight Status Among Girls Who Are 'Picky Eaters,'" *Journal of the American Dietetic Association* 105 4 (2005): 541–48.

Kerzner, Benny, Kim Milano, William C. Maclean, Glenn B. Berall, Sheela Stuart, and Irene Chatoor. "A Practical Approach to Classifying and Managing Feeding Difficulties." *Pediatrics* 135 2 (2015): 344–53.

Mazzeo, Suzanne E., Nancy Zucker, Clarice K. Gerke, Karen S. Mitchell, and Cynthia M. Bulik. "Parenting Concerns of Women with Histories of Eating Disorders." *International Journal of Eating Disorders* 37 Suppl. (2005): S77–9; discussion S87–9.

Zickgraf, Hana F., and Katie Schepps. "Fruit and Vegetable Intake and Dietary Variety in Adult Picky Eaters," *Food Quality and Preference* 54 (2016): 39–50.

Zucker, Nancy, William C. Copeland, Lauren Franz, Kimberly Carpenter, Lori A. Keeling, Adrian Angold, and Helen Link Egger. "Psychological and Psychosocial Impairment in Preschoolers with Selective Eating." *Pediatrics* 137 1 (2015).

The Picky Eaters Association Facebook Page: https://www.facebook.com/groups/55498983024/

THE DEFINITION OF AVOIDANT-RESTRICTIVE FOOD INTAKE DISORDER (ARFID) PER:

American Psychiatric Association. "Feeding and Eating Disorders," *Diagnostic and Statistical Manual of Mental Disorders, Fifth Edition*. Arlington, VA: American Psychiatric Publishing, 2013.

71 PERCENT OF CHILDREN WITH FEEDING PROBLEMS RELY ON PEDIASURE OR FEEDING TUBE SUPPLEMENTATION, PER:

Williams, Keith E., Helen M. Hendy, Douglas G. Field, Yekaterina Belousov, Katherine Riegel, and Whitney Harclerode. "Implications of Avoidant/Restrictive Food Intake Disorder (ARFID) on Children with Feeding Problems." *Children's Health Care* 44 4 (2015): 307–21.

STATISTICS ON FREQUENCY AND COMPLICATION RATES OF ARFID AND OTHER EATING DISORDERS FROM:

Fisher, Martin M., David A. S. Rosen, Rollyn M. Ornstein, Kathleen A. Mammel, Debra K. Katzman, Ellen S. Rome, S. Todd Callahan, Joan B. Malizio, Sarah Kearney, and B. Timothy Walsh. "Characteristics of Avoidant/Restrictive Food Intake Disorder in Children and Adolescents: A 'New Disorder' in DSM-5." *Journal of Adolescent Health* 55 1 (2014): 49–52.

Forman, Sara F., N. Ryan McKenzie, Rebecca Hehn, M. Monge, Cynthia J. Kapphahn, Kathleen A. Mammel, S. Todd Callahan, Eric J. Sigel, Terrill D. Bravender, Mary Romano, Ellen S. Rome, Kelly A. Robinson, Martin Fisher, Joan B. Malizio, David A. S. Rosen, Albert C. Hergenroeder, Sara M. Buckelew, M. Susan Jay, Jeffrey Lindenbaum, Vaughn I. Rickert, Andrea K. Garber, Neville H. Golden, and Elizabeth Woods. "Predictors of Outcome at 1 Year in Adolescents with DSM-5 Restrictive Eating Disorders: Report of the National Eating Disorders Quality Improvement Collaborative." *Journal of Adolescent Health* 55 6 (2014): 750–6.

Hoek, Hans Wijbrand. "Incidence, Prevalence and Mortality of Anorexia Nervosa and Other Eating Disorders." *Current Opinion in Psychiatry* 19 4 (2006): 389–94.

Kurz, Susanne, Zoé van Dyck, Daniela Dremmel, Simone Munsch, and Anja Hilbert. "Early-Onset Restrictive Eating Disturbances in Primary School Boys and Girls." *European Child and Adolescent Psychiatry* (2014).

Nicely, Terri A., Susan E. Lane-Loney, Emily Masciulli, Christopher S. Hollenbeak, and Rollyn M. Ornstein. "Prevalence and Characteristics of Avoidant/Restrictive Food Intake Disorder in a Cohort of Young Patients in Day Treatment for Eating Disorders." *International Journal of Eating Disorders* (2014).

Norris, Mark L., and Debra K. Katzman. "Change Is Never Easy, but It Is Possible: Reflections on Avoidant/Restrictive Food Intake Disorder Two Years After Its Introduction in the DSM-5." *Journal of Adolescent Health* 57 1 (2015): 8–9.

Norris, Mark L., Amy Robinson, Nicole Obeid, Megan Harrison, Wendy J. Spettigue, and Katherine Henderson. "Exploring Avoidant/ Restrictive Food Intake Disorders in Eating Disorder Patients: A Descriptive Study." *International Journal of Eating Disorders* 47 5 (2014): 495–99.

Norris, Mark L., Wendy J. Spettigue, and Debra K. Katzman. "Update on Eating Disorders: Current Perspectives on Avoidant/Restrictive Food Intake Disorder in Children and Youth." *Neuropsychiatric Disease and Treatment* (2016).

Ornstein, Rollyn M., David A. S. Rosen, Kathleen A. Mammel, S. Todd Callahan, Sara Forman, M. Susan Jay, Martin Fisher, Ellen Rome, and B. Timothy Walsh. "Distribution of Eating Disorders in Children and Adolescents Using the Proposed DSM-5 Criteria for Feeding and Eating Disorders." *Journal of Adolescent Health* 53 2 (2013): 303–5.

Satter, Ellyn. "ARFID: What Is It? What Does It Have to Do with Feeding Dynamics and Eating Competence?" *Family Meals Focus* (newsletter) No. 89, August 2015.

Strandjord, Sarah Elizabeth, Erin H. Sieke, Miranda Richmond, and Ellen S. Rome. "Avoidant/Restrictive Food Intake Disorder: Illness and Hospital Course in Patients Hospitalized for Nutritional Insufficiency." *Journal of Adolescent Health* 57 6 (2015): 673–8.

RESEARCH ON THE DEVELOPMENT OF PICKY EATING:

Hayes, John E., and Russell S. J. Keast. "Two Decades of Supertasting: Where Do We Stand?" *Physiology & Behavior* 104 5 (2011): 1072–74.

Kauer, Jane, Marcia Levin Pelchat, Paul Rozin, and Hana F. Zickgraf. "Adult Picky Eating: Phenomenology, Taste Sensitivity, and Psychological Correlates." *Appetite* 90 (2015): 219–28.

Rowell, Katja, and Jenny McGlothlin. *Helping Your Child with Extreme Picky Eating: A Step-by-Step Guide for Overcoming Selective Eating, Food Aversion and Feeding Disorders.* Oakland, CA: New Harbinger Publications, May 2015.

ON FAMILY-BASED THERAPY (FBT) AS A TREATMENT FOR ANOREXIA:

Brown, Harriet. *Brave Girl Eating: A Family's Struggle with Anorexia.* New York: William Morrow, 2011.

RESEARCH ON THE LINK BETWEEN ANOREXIA AND PICKY EATING:

Dellava, Jocilyn E., Sara E. Trace, Michael Strober, Laura M. Thornton, Kelly L. Klump, Harry Brandt, Steve Crawford, Manfred M. Fichter, Katherine A. Halmi, Craig Johnson, Allan S. Kaplan, James E. Mitchell, Janet Treasure, D. Blake Woodside, Wade H. Berrettini, Walter H. Kaye, and Cynthia M. Bulik. "Retrospective Maternal Report of Early Eating Behaviours in Anorexia Nervosa." *European Eating Disorders Review* 20 2 (2012): 111–15.

Micali, Nadia, Jason A. Holliday, Andreas Karwautz, Maria Haidvogl, G. Wagner, Fernando Fernández-Aranda, Antonella Badia, Luciana V. Gimenez, Raquel Solano, Maria Brecelj-Anderluh, R. Mohan, D. S. Collier, and Jonathan Treasure. "Childhood Eating and Weight in Eating Disorders: A Multi-Centre European Study of Affected Women and Their Unaffected Sisters." *Psychotherapy and Psychosomatics* 76 4 (2007): 234–41.

RESEARCH ON TREATMENT FOR ARFID:

Pizzo, Bianca, Molly Coyle, Laura Seiverling, and Keith Williams. "Plate A–Plate B: Use of Sequential Presentation in the Treatment of Food Selectivity." *Behavioral Interventions* 27 4 (2012): 175–84.

Williams, Keith E., Douglas G. Field, Katherine Riegel, and Candace Paul. "Brief, Intensive Behavioral Treatment of Food Refusal Secondary to Emetophobia." *Clinical Case Studies* 10 4 (2011): 304–11.

Zickgraf, Hana F., Martin E. Franklin, and Paul Rozin. "Adult Picky Eaters with Symptoms of Avoidant/Restrictive Food Intake Disorder: Comparable Distress and Comorbidity but Different Eating Behaviors Compared to Those with Disordered Eating Symptoms." *Journal of Eating Disorders* (2016).

Zickgraf, Hana F., and Katie Schepps. "Fruit and Vegetable Intake and Dietary Variety in Adult Picky Eaters." *Food Quality and Preference* 54 (2016): 39–50.

CHAPTER FIVE

THIS CHAPTER WAS INFORMED BY SOME OF MY PREVIOUS WRITING:

Sole-Smith, Virginia. "Getting Jobbed." *Harper's*, October 2015.

———. "The Hungry House." *Parents*, July 2011.

THE FOLLOWING SERVED AS GENERAL SUBJECT MATTER RESEARCH FOR THIS CHAPTER:

Berg, Joel. *All You Can Eat: How Hungry Is America?* New York: Seven Stories Press, 2008.

Bittman, Mark. *Food Matters: A Guide to Conscious Eating with More Than 75 Recipes.* New York: Simon & Schuster, 2008.

Gustafson, Kaaryn. *Cheating Welfare: Public Assistance and the Criminalization of Poverty.* New York: New York University Press, 2011.

Guthman, Julie. *Weighing In: Obesity, Food Justice and the Limits of Capitalism.* Berkeley: University of California Press, 2011.

Hays, Sharon. *Flat Broke with Children: Women in the Age of Welfare Reform.* New York: Oxford University Press, 2004.

Irby, Samantha. *We Are Never Meeting in Real Life: Essays.* New York: Vintage, 2017.

Nestle, Marion. *Food Politics: How the Food Industry Influences Nutrition and Health.* Berkeley: University of California Press, 2013.

Pollan, Michael. *Food Rules: An Eater's Manual.* New York: Penguin, 2009.

————. *The Omnivore's Dilemma: A Natural History of Four Meals.* New York: Penguin, 2007.

————. *What to Eat.* New York: North Point Press, 2007.

Poppendieck, Janet. *Free for All: Fixing School Food in America.* Berkeley: University of California Press, 2010.

Zucchino, David. *Myth of the Welfare Queen.* New York: Simon & Schuster, 1999.

GROWTH OF FARMERS MARKETS IN THE UNITED STATES:

Ragland, Edward, and Debra Tropp. *USDA National Farmers Market Manager Survey, 2006.* U.S. Department of Agriculture, Agricultural Marketing Service, May 2009.

U.S. Department of Agriculture, Agricultural Marketing Service. *New Data Reflects Continued Demand for Farmers Markets* (press release), August 4, 2014.

U.S. Department of Agriculture, Agricultural Marketing Service. *USDA Awards Grants to Boost Access to Farmers Markets, Nutritious Food for SNAP Participants* (press release), September 9, 2016.

THE RISE OF DIETING THEMES IN ALTERNATIVE-FOOD MOVEMENT:

Parker-Pope, Tara. "Michael Pollan Offers 65 Ways to Eat Food." *Well* (blog). *New York Times,* January 28, 2010. https://well.blogs.nytimes.com /2010/01/08/michael-pollan-offers-64-ways-to-eat-food/.

STATISTICS ON POVERTY, FOOD INSECURITY, AND RACE:

Monsivais, P., and A. Drewnowski. "The Rising Cost of Low-Energy-Density Foods." *Journal of the American Dietetic Association* 107 12 (2007): 2071–6.

Feeding America. "Hunger and Poverty Fact Sheet," May 2017.

————. "African American Hunger Facts," May 2017.

————. "Child Hunger Fact Sheet," March 2017.

USDA. "Definitions of Food Security," published by the Economic Research Service.

USDA. "Access to Affordable and Nutritious Food: Measuring and Understanding Food Deserts and Their Consequences." Report to Congress. June 2009.

THE ALTERNATIVE-FOOD MOVEMENT'S APPROPRIATION OF
BLACK FOOD CULTURE:

Lewis, Edna. *The Taste of Country Cooking: 30th Edition*. New York: Knopf, 2006.

Simley, Shakirah. "A More Abundant Share: The Future of Food Is Black." *Huffington Post*, February 4, 2017.

FOOD MARKETING IN MARGINALIZED COMMUNITIES:

DiSantis, Katherine Isselmann, Amy E. Hillier, Rio Holaday, and Shiriki K. Kumanyika. "Why Do You Shop There? A Mixed Methods Study Mapping Household Food Shopping Patterns onto Weekly Routines of Black Women." *International Journal of Behavioral Nutrition and Physical Activity* (2016).

Harris, Jennifer L., Catherine Shehan, Renee Gross, Shiriki Kumanyika, Vikki Lassiter, Amelie G. Ramirez, and Kipling Gallion. "Rudd Report: Food Advertising Targeted to Hispanic and Black Youth: Contributing to Health Disparities." University of Connecticut Rudd Center for Food Policy & Obesity, August 2015.

Kumanyika, Shiriki. "Beverage Marketing as Public Health Policy Target." *American Journal of Public Health* (2015).

Kuzemchak, Sally. "How the LA Unified School District Shut Down a McDonald's Fundraiser." *Civil Eats*, May 12, 2017.

Phipps, E., Shiriki K. Kumanyika, Shana D. Stites, S. Brook Singletary, Clarissa A. Cooblall, and Katherine Isselmann DiSantis. "Buying Food on Sale: A Mixed Methods Study with Shoppers at an Urban Supermarket, Philadelphia, Pennsylvania 2010–2012." *Preventing Chronic Disease* (2014).

Powell, Lisa M., Roy Wada, and Shiriki Kumanyika. "Racial/Ethnic and Income Disparities in Child and Adolescent Exposure to Food and Beverage Television Ads Across US Media Markets." *Health & Place* 29 (2014): 124–31.

Wilcox, Brian L., Dale Kunkel, Joanne Cantor, Peter Dowrick, Susan Linn, and Edward Palmer. "Report of the APA Task Force on Advertising and Children." American Psychological Association, February 20, 2004.

ROLE OF WEIGHT, FOOD, AND HEALTH IN MARGINALIZED COMMUNITIES:

Jain, Anjali, Susan N. Sherman, Leigh A. Chamberlin, Yvette Carter, Scott W. Powers, and Robert C. Whitaker. "Why Don't Low-Income Mothers Worry About Their Preschoolers Being Overweight?" *Pediatrics* 107 5 (2001).

Kumanyika, Shiriki, Morgan L. Swank, James J. Stachecki, Melicia C. Whitt-Glover, and Louise Brennan. "Examining the Evidence for Policy and Environmental Strategies to Prevent Childhood Obesity in Black Communities: New Directions and Next Steps." *Obesity Reviews* 15 Suppl. 4 (2014): 177–203.

Parks, Elizabeth Prout, Anne Kazak, Shiriki K. Kumanyika, Lisa Scandale Lewis, and Frances K. Barg. "Perspectives on Stress, Parenting, and Children's Obesity-Related Behaviors in Black Families." *Health Education & Behavior* 43 6 (2016): 632–40.

Smith, Jaime, Kari Adamsons, Rachel Vollmer, and Amy Mobley. "Is Nutrition Assistance Program Participation or Food Security Status Associated with Parental Feeding Style or Child Body Mass Index?" *FASEB Journal* 29 Suppl. 1 (2015).

Vollmer, Rachel, and Amy Mobley. "A Pilot Study to Explore How Low-Income Mothers of Different Ethnic/Racial Backgrounds Perceive and Implement Recommended Childhood Obesity Prevention Messages." *Childhood Obesity* 9 3 (2013).

Vollmer, Rachel, Kari Adamsons, Jaime Foster, and Amy Mobley. "Do Low-Income Mothers and Fathers of Preschool Children Share Similar Concerns and Perception About Their Child's Body Weight?" *FASEB Journal* 29 Suppl. 1 (2015).

Whitt-Glover, Melicia C., Shiriki Kumanyika, and Debra L. Haire-Joshu. "Introduction to the Special Issue on Achieving Healthy Weight in Black American Communities." *Obesity Reviews* 15 Suppl. 4 (2014): 1–4.

FOOD ACTIVISM IN POOR COMMUNITIES:

Bittman, Mark, Michael Pollan, Olivier De Schutter, and Ricardo Salvador. "Food and More: Expanding the Movement for the Trump Era." *Civil Eats,* January 16, 2017.

Guthman, Julie. "Can't Stomach It: How Michael Pollan et al. Made Me Want to Eat Cheetos." *Gastronomica: The Journal of Critical Food Studies* 7 3 (2007): 75–79.

McMillan, Tracie. "Food's Class Warfare: Do Poor People Eat Badly Because of Limited Options or Personal Preference?" *Slate,* June 27, 2012.

O'Connor, Anahad. "In the Shopping Cart of a Food Stamp Household: Lots of Soda." *New York Times,* January 13, 2017.

Oyston, Glenys. "Food Is the New Classism." *Health at Every Size Blog,* May 9, 2017. https://healthateverysizeblog.org/2017/05/09/the-haes-files -food-is-the-new-classism/.

Philpott, Tom. "Why Michael Pollan and Alice Waters Should Quit Celebrating Food-Price Hikes." *Grist,* April 5, 2008.

Pollan, Michael. "An Open Letter to the Next Farmer-in-Chief." *The New York Times Magazine,* October 9, 2008.

Vallas, Rebecca, and Katherine Gallagher Robbins. "In the Shopping Cart of a Food Stamp Household: Not What the *New York Times* Reported."

TalkPoverty.org, January 16, 2017. https://talkpoverty.org/2017/01/16 /shopping-cart-food-stamp-household-not-new-york-times-reported/.

USDA. "Food Typically Purchased by Supplemental Nutrition Assistance Program (SNAP) Households," Nutrition Assistance Program Report, November 2016.

CHAPTER SIX

THIS CHAPTER WAS INFORMED BY SOME OF MY PREVIOUS WRITING:

Sole-Smith, Virginia. "Can You Be Heavy and Healthy?" *Marie Claire*, August 2014.

————. "The Energizer Body." *Real Simple*, February 2015.

————. *Never Say Diet*, a body image blog published by iVillage.com in 2011. Archived at http://virginiasolesmith.com/category/special-projects /never-say-diet/.

————. "What Happens When Your Body Image Nightmare Comes True?" *Buzzfeed*, March 23, 2015.

THE FOLLOWING SERVED AS GENERAL SUBJECT MATTER RESEARCH FOR THIS CHAPTER:

Bacon, Linda. *Health at Every Size: The Surprising Truth About Your Weight*. Dallas: BenBella Books, 2010.

Bacon, Linda, and Lucy Aphramor. *Body Respect: What Conventional Health Books Get Wrong, Leave Out, and Just Plain Fail to Understand About Weight*. Dallas: BenBella Books, 2014.

Brown, Harriet. *Body of Truth: How Science, History and Culture Drive Our Weight—and What We Can Do About It*. New York: Da Capo, 2016.

Chastain, Ragen. *FAT: The Owner's Manual*. Austin, TX: Sized for Success Multimedia, 2012.

Gay, Roxane. *Hunger: A Memoir of (My) Body*. New York: Harper, 2017.

Kirkland, Anna. *Fat Rights: Dilemmas of Difference and Personhood.* New York: New York University Press, 2008.

Kessler, David. *The End of Overeating: Taking Control of the Insatiable American Appetite.* New York: Rodale, 2009.

Kolata, Gina. *Rethinking Thin: The New Science of Weight Loss—and the Myths and Realities of Dieting.* New York: Farrar, Straus and Giroux, 2007.

Mann, Traci. *Secrets from the Eating Lab: The Science of Weight Loss, the Myth of Willpower, and Why You Should Never Diet Again.* New York: Harper Wave, 2017.

Weiner, Jennifer. *Hungry Heart: Adventures in Life, Love, and Writing.* New York: Atria, 2016.

West, Lindy. *Shrill: Notes from a Loud Woman.* New York: Hachette, 2016.

DATA ON BARIATRIC SURGERY:

American Society for Metabolic and Bariatric Surgery. "Estimate of Bariatric Surgery Numbers, 2011–2015," July 2016; "Metabolic and Bariatric Surgery Fact Sheet," November 2013; "Bariatric Surgery Procedures." https://asmbs.org/.

Boyd, Owen F., and N. J. Jackson. "How Is Risk Defined in High-Risk Surgical Patient Management?" *Critical Care* 9 (2005): 390–96.

Inaba, Colette S., Christina Y. Koh, Sarath Sujatha-Bhaskar, Jack P. Silva, Yanjun Chen, Danh V. Nguyen, and Ninh T. Nguyen. "One-Year Mortality After Contemporary Laparoscopic Bariatric Surgery: An Analysis of the Bariatric Outcomes Longitudinal Database." *Journal of the American College of Surgeons* (2018).

Kolata, Gina. "A Year After Weight-Loss Surgery: A Year of Joys and Disappointments." *New York Times*, December 27, 2016.

Omalu, Bennet I., Diane G. Ives, Alhaji M. Buhari, Jennifer L. Lindner, Philip R. Schauer, Cyril H. Wecht, and Lewis H. Kuller. "Death Rates and Causes of Death After Bariatric Surgery for Pennsylvania Residents, 1995 to 2004." *Archives of Surgery* 142 10 (2007): 923–28; discussion 929.

Ponce, Jaime. "The Numbers Are In: Sleeve Gastrectomy Remains Top Procedure." *Connect: The Official News Magazine of ASMBS*, August 2016.

RESEARCH ON OBESITY, WEIGHT LOSS EFFORTS, AND DIETING
SUCCESS RATES:

American Society of Metabolic and Bariatric Surgery and NORC at the University of Chicago. *Obesity Survey*, 2016.

Atallah, Renée, Kristian B. Filion, Susan M. Wakil, Jacques Genest, Lawrence Joseph, Paul Poirier, Stéphane Rinfret, Ernesto L. Schiffrin, and Mark J. Eisenberg. "Long-Term Effects of 4 Popular Diets on Weight Loss and Cardiovascular Risk Factors." *Circulation: Cardiovascular Quality and Outcomes* 7 (2014): 815–27.

Brodesser-Akner, Taffy. "Losing It in the Anti-Diet Age." *The New York Times Magazine*, August 2, 2017.

Hudson, James I., Eva Hiripi, Harrison Graham Pope, and Ronald C. Kessler. "The Prevalence and Correlates of Eating Disorders in the National Comorbidity Survey Replication." *Biological Psychiatry* 61 3 (2007): 348–58.

Lowe, Michael R., Tanja V. E. Kral, and Karen Miller-Kovach. "Weight-Loss Maintenance 1, 2 and 5 Years After Successful Completion of a Weight-Loss Programme." *British Journal of Nutrition* 99 4 (2008): 925–30.

Katz, David. *The Flavor Point Diet*. New York: Rodale, 2005.

Katz, David L., and Stephan Meller. "Can We Say What Diet Is Best for Health?" *Annual Review of Public Health* 35 (2014): 83–103.

Mann, Traci L., A. Janet Tomiyama, Erika H. Westling, A. J. Lew, Barbra Samuels, and Jason Chatman. "Medicare's Search for Effective Obesity Treatments: Diets Are Not the Answer." *American Psychologist* 62 3 (2007): 220–33.

Mann, Traci. "Oprah's Investment in Weight Watchers Was Smart Because the Program Doesn't Work." *New York*, October 2015.

Wallis, Lucy. "Do Slimming Clubs Work?" BBC, August 8, 2013. http://www.bbc.com/news/magazine-23463006/.

THE WILLPOWER MYTH:

Luck-Sikorski, Claudia, S. G. Riedel-Heller, and Jo C. Phelan. "Changing Attitudes Towards Obesity—Results from a Survey Experiment." *BMC Public Health* (2017).

Kolata, Gina. "Americans Blame Obesity on Willpower Despite Evidence It's Genetic." *New York Times*, November 1, 2016.

RESEARCH ON BARIATRIC SURGERY COMPLICATIONS AND SIDE EFFECTS:

Bacon, Linda. "A Message for People Considering Bariatric Surgery" and "Dreams on the Operating Room Table (Bariatric Surgery)" in *Health at Every Size: The Surprising Truth About Your Weight*. Dallas: BenBella Books, 2010.

Conceição, Eva, Ana Vaz, Ana Pinto Bastos, Ana Ramos, and Paulo Machado. "The Development of Eating Disorders After Bariatric Surgery." *Eating Disorders* 21 3 (2013): 275–82.

Conceição, Eva, Molly Orcutt, James Mitchell, Scott Engel, Kim Lahaise, Michelle Jorgensen, Kara Woodbury, Naomi Hass, Luis García, and Stephen A. Wonderlich. "Eating Disorders After Bariatric Surgery: A Case Series." *International Journal of Eating Disorders* 46 3 (2013): 274–79.

Driscoll, Shannon, Deborah M. Gregory, John M. Fardy, and Laurie K. Twells. "Long-Term Health-Related Quality of Life in Bariatric Surgery Patients: A Systematic Review and Meta-Analysis." *Obesity* 24 1 (2016): 60–70.

Hillstrom, Kathryn, and Nicole M. Avila. "BED, Bulimia in Bariatric Surgery Patients." *Today's Dietitian* 16 1 (2014): 12.

Van Beek, Adriana P. A., Marloes Emous, Marie Paule Laville, and Jesse Tack. "Dumping Syndrome After Esophageal, Gastric or Bariatric Surgery: Pathophysiology, Diagnosis, and Management." *Obesity Reviews* 18 1 (2017): 68–85.

RESEARCH SUPPORTING THE HEALTH AT EVERY SIZE CASE:

Bacon, Linda C., and Lucy Aphramor. "Weight Science: Evaluating the Evidence for a Paradigm Shift." *Nutrition Journal* 10 9 (2011).

Chaput, J. Scott, Zachary Michael Ferraro, Denis Prud'homme, and Arya M. Sharma. "Widespread Misconceptions About Obesity." *Canadian Family Physician* 60 11 (2014): 973–75, 981–84.

Tomiyama, A. Janet, Britt Ahlstrom, and Traci Mann. "Long-Term Effects of Dieting: Is Weight Loss Related to Health?" *Social and Personality Psychology Compass* 7 12 (2013): 861–77.

RESEARCH ON WEIGHT BIAS AND STIGMA:

Brochu, Paula M., Rebecca L. Pearl, Rebecca M. Puhl, and Kelly D. Brownell. "Do Media Portrayals of Obesity Influence Support for Weight-Related Medical Policy?" *Health Psychology* 33 2 (2014): 197–200.

DuBreuil, Lisa. "Weight Stigma and Weight Loss Surgery." *Binge Eating Disorder Association* (blog), September 22, 2013. https://bedaonline.com /wsaw2013/health-weight-stigma-sizes-lisa-dubreuil-licsw/.

Hunger, Jeffrey M., Brenda N. Major, Alison Blodorn, and Carol T. Miller. "Weighed Down by Stigma: How Weight-Based Social Identity Threat Contributes to Weight Gain and Poor Health." *Social and Personality Psychology Compass* 9 6 (2015): 255–268.

Puhl, Rebecca M., and Chelsea A. Heuer. "Obesity Stigma: Important Considerations for Public Health." *American Journal of Public Health* 100 6 (2010): 1019–28.

Puhl, Rebecca M., and Young Suh. "Health Consequences of Weight Stigma: Implications for Obesity Prevention and Treatment." *Current Obesity Reports* 4 2 (2015): 182–190.

Schwartz, Marlene B., Heather O'Neal Chambliss, Kelly D. Brownell, Steven Noel Blair, and Charles Billington. "Weight Bias Among Health Professionals Specializing in Obesity." *Obesity Research* 11 9 (2003): 1033–39.

Tomiyama, A. Janet, and Traci Mann. "If Shaming Reduced Obesity, There Would Be No Fat People." *Hastings Center Report* 43 3 (2013).

Wellman, Joseph D., Ashley M. Araiza, Ellen E. Newell, and Shannon K. McCoy. "Weight Stigma Facilitates Unhealthy Eating and Weight Gain via Fear of Fat." *Stigma and Health* (2017).

RESEARCH ON FOOD ADDICTION AND LINK BETWEEN WEIGHT-LOSS SURGERY AND ALCOHOLISM:

DuBreuil, Lisa, and Stephanie Sogg. "Alcohol-Use Disorders After Bariatric Surgery: The Case for Targeted Group Therapy." *Current Psychiatry* 16 1 (2017): 39–44, 48–49.

Gearhardt, Ashley N., William R. Corbin, and Kelly D. Brownell. "Preliminary Validation of the Yale Food Addiction Scale." *Appetite* 52 2 (2009): 430–36.

Sharma, Arya. "Why Does the Food Addiction Model of Obesity Management Lack Good Science?" *Dr. Sharma's Obesity Notes* (blog), January 26, 2016. http://www.drsharma.ca/why-does-the-food-addiction-model-of -obesity-management-lack-good-science/.

Tordoff, Michael G., Jordan A. Pearson, Hillary T. Ellis, and Rachel L. F. Poole. "Does Eating Good-Tasting Food Influence Body Weight?" *Physiology & Behavior* 170 (2017): 27–31.

CHAPTER SEVEN

Images from Lois Bielefeld's "Weeknight Dinners" series can be found at loisbielefeld.com.

RESEARCH ON FOOD AND MORALITY:

Casey, Rebecca, and Paul Rozin. "Changing Children's Food Preferences: Parent Opinions." *Appetite* 12 3 (1989): 171–82.

Geier, Andrew B., and Paul Rozin. "Weighing Discomfort in College Age American Females: Incidence and Causes." *Appetite* 51 1 (2008): 173–77.

Hormes, Julia M., Paul Rozin, Melanie C. Green, and Katrina Fincher. "Reading a Book Can Change Your Mind, but Only Some Changes Last for a Year: Food Attitude Changes in Readers of *The Omnivore's Dilemma.*" *Frontiers in Psychology* 4 (2013): 778.

Rozin, Paul. "The Meaning of Food in Our Lives: A Cross-Cultural Perspective on Eating and Wellbeing." *Journal of Nutrition Education and Behavior* 37 2 (2005): 107–12.

Rozin, Paul, Rebecca Bauer, and Dana Catanese. "Food and Life, Pleasure and Worry, Among American College Students: Gender Differences and Regional Similarities." *Journal of Personality and Social Psychology* 85 1 (2003): 132–41.

Rozin, Paul, Nicole Kurzer, and Adam B. Cohen. "Free Associations to 'Food': The Effects of Gender, Generation and Culture." *Journal of Research in Personality* 36 5 (2002): 419–41.

Rozin, Paul, Shlomo Trachtenberg, and Adam B. Cohen. "Stability of Body Image and Body Image Dissatisfaction in American College Students over About the Last 15 Years." *Appetite* 37 3 (2001): 245–8.

RESEARCH ON ABBY SOLOMON'S CONDITION:

Kennedy, Pagan. "The Thin Gene." *New York Times*, November 25, 2016.

Romere, Chase, Clemens Duerrschmid, Juan Bournat, Petra Constable, Mahim Jain, Fan Xia, Pradip Kr Saha, Maria Del Solar, Bokai Zhu, Brian J. York, Poonam Sarkar, David A. Rendon, M. Waleed Gaber, Scott A. Lemaire, Joseph S. Coselli, Dianna McGookey Milewicz, Vernon Reid Sutton, Nancy F. Butte, David D. Moore, and Atul R. Chopra. "Asprosin, a Fasting-Induced Glucogenic Protein Hormone." *Cell* 165 (2016): 566–79.

WEIGHT WATCHERS' TUMBLING SALES AND OUR CULTURAL SHIFT AWAY FROM DIETING:

Brodesser-Akner, Taffy. "Losing It in the Anti-Diet Age." *The New York Times Magazine*, August 2, 2017.

RESEARCH ON POST-DIETING WEIGHT REGAIN:

Mann, Traci L., A. Janet Tomiyama, Erika H. Westling, A. J. Lew, Barbra Samuels, and Jason Chatman. "Medicare's Search for Effective Obesity Treatments: Diets Are Not the Answer." *American Psychologist* 62 3 (2007): 220–33.

ACKNOWLEDGMENTS

An ongoing challenge of this book was that I had to live entire chapters before I could write them. I am indebted to everyone who supported me through both parts of the process.

My incomparable agent, Jin Auh, made this book better every time we talked about it. My editor, Serena Jones, embraced it straightaway and provided many keen insights (plus useful baby sleep-training tips). My fact-checker, Tova Carlin, took tremendous care with the manuscript and with my sources. My talented research assistant, Allison Pohle, cheerfully rose to every challenge, no matter how obscure the request. Gabrielle Gerard made taking author photos more fun than terrifying. Thank you as well to Madeline Jones, Hannah Campbell, Jason Liebman, Carolyn O'Keefe, Jessica Wiener, Maggie Richards, and everyone at Henry Holt who shepherded this book along.

Many friends and colleagues read bits of this book in varying states of acceptability and I so appreciate their generosity of both time and talent. My neighborhood writing group—Lauren Daisley, Liesa Goins, Melinda Wenner Moyer, Maria Ricapito, and Nicki Sizemore—provided smart critiques, support, and snacks. My terrific editors, including Danielle Claro, Kara Corridan, Diane Debrovner, Barbara Ehrenreich, Noelle Howey, Esther Kaplan, Gail O'Connor, Alex Postman, Alissa Quart, and Genevieve Smith, all collaborated with me on stories that either directly informed a chunk of this project or taught me something new and indispensable about reporting and story craft. I am especially grateful to Dean Robinson and Claire Gutierrez of *The New York Times Magazine* for helping me tell the story of how Violet learned to eat again, which led me to every other story in this book. And, a profound thank-you to my first boss, mentor, and dear friend, Jennifer Braunschweiger, who, when I was burying myself in research and panic, told me: "Just start writing." She was right, as always.

One question I am asked far too often (and male authors aren't asked enough) is how I balance writing and parenting. The answer: childcare. Thank you to all the devoted babysitters and preschool teachers who have loved and nurtured our girls when we're working, particularly Sarina Cass, Erica Davis, Sherri Oakley, Beth Flanagin, Larissa Nordone, and Karen DeRosa. While we're on the subject of education, I'll also mention my high school English teacher Ilona Fucci, because everyone has that one special teacher, and she was mine.

I must also express my gratitude for Violet's medical team, not only for providing her with meticulous care but also for educating me on issues that provided important jumping-off points for my research and thinking in these pages. We can never adequately thank Joseph Giamelli, M.D., Christa Miliaresis, M.D., Michelle Patrick, N.P., Gary Tatz, M.D., or Suvro Sett, M.D., because even writers can't find words for the people who saved their daughter's life. The same goes for Samir Pandya, M.D., the surgeon who placed the G-tube that enabled Violet to thrive on tube feeds and later performed several operations to save her lungs. We also adore Tricia Hiller, Jean Lavin, Ana Riccio, and all the nurses in the pediatric intensive care unit at Maria Fareri Children's Hospital at Westchester Medical Center in Valhalla, New York, who did so much for our family. And of course, Lynne Cross Menard and Maggie Ruzzi taught us how to teach Violet to eat.

An entire community, online and off, has followed Violet's medical progress and I am grateful to everyone who supported us during the dark times. Katherine Nolan Brown, Amy Palanjian, Liz Scranton, and Kate Tellers, my best friends since we were barely grown-ups (and in Liz's case, long before), were with me almost hourly in one way or another on the worst and best days. As my longtime work wife, Amy was also involved in every step of this project and was the first friend to brave reading through the finished manuscript. Thank you.

Kathy and Steve Upham have been my third set of parents since I was a teenager. Becca Huben, Dan Huben, Matt Upham, and Sara Boyorak are the kind of siblings-in-law who are also dear friends. Their children (Lorelai Huben, Nate Huben, and Douglass Upham) are a constant delight. I'm also thankful for my many remarkable grandparents, aunts, uncles, and cousins on both sides of the Atlantic, and especially for the other writers in the family: My late step-grandmother, U. T. Miller Summers, and my cousins Chandler Klang Smith and Kate Summers, who are constant inspirations.

Two of the smartest and kindest humans on the planet are my siblings, Caroline Smith and Reed Smith, and I'm lucky that they always have my back. We were raised by four extraordinary parents who encouraged us to chase after every dream: My father, Rogers Smith, who gave me my work ethic and the knowledge that with great power comes great responsibility. My stepmother, Mary Summers, who is the reason Chapter 5 exists. My stepfather, Pat Russo, who taught me to question everything. And my mother, Marian Sole, who knew I was a writer before I did (and then made sure I figured it out).

But most essential to the writing of this book were these three: My sweet baby Beatrix, who obligingly let me meet my manuscript deadline before she made her entrance into the world. My mighty Violet, who has always known herself so well and has taught me so much—and who, I hope, will be proud that I chose to tell this little bit of her story. And Dan, my best friend and my home for twenty years. Thank you for climbing these mountains with me. I love you.

INDEX

ABOUT THE AUTHOR

As a journalist, VIRGINIA SOLE-SMITH has reported from kitchen tables and grocery stores, graduated from beauty school, and gone swimming in a mermaid's tail. Her work has appeared in *The New York Times Magazine*, *Harper's*, *Elle*, and many other publications. She is also a contributing editor with *Parents* magazine. The Nation Institute's Investigative Fund and the Economic Hardship Reporting Project have supported several of her stories. Sole-Smith lives with her husband and two daughters in the Hudson Valley. Find her at virginiasolesmith.com and @v_solesmith on social media.